The Nerd Whisperer

David A. Oppenheimer

Cover Design By:
Beetiful Book Covers (www.beetiful.com)

DEDICATION

This book is dedicated to all the men and women who create the software that runs our daily lives. Through their hard work, dedication, attention to detail, and creativity, they make our world a better place.

ACKNOWLEDGMENTS

I would like to acknowledge the encouragement, support, and brilliant minds of Dr. John Bordeaux, Richard Hildebrand, Hong Kim, Troy Pomroy, and John Wood who listened to me ramble about management topics and always brought me back home to the salient points.

I thank all of the mentors I've had over the years, and there have been many. In particular I'd like to thank Alfred G. "Waboos" Hare, Stanley Lucas, Ed McGushin, Roland Fisher, Dr. Ernst Volgenau, Barbara Sada, Dr. Hatte Blejer, and Katharine Murphy for hiring me, challenging me, growing me, and most of all putting up with me.

I thank my parents for always supporting my crazy endeavors even when they didn't fully understand what I do for a living. I thank my mother-in-law, Lee Ann Draud, for making time in a schedule with no time at all to advise, review, and edit this mighty tome. I thank my daughter for putting up with years of listening to stories about projects in which she could not reasonably be expected to have any interest whatsoever, and for doing it with such grace.

Lastly and firstly I'd like to thank my wife, Elise, for supporting me throughout this effort and throughout our lives together. She supports me in all ways: emotionally, financially, intellectually, and nutritionally. She is my constant sounding board, editor, coach, motivator, and voice of reason. I love you.

Introduction

What I hope to do here is to explain the nature and motivations of the people behind the technology and how those motivations, skills, and personalities interact when you are trying to get a project accomplished. Having a better understanding of what drives these people will allow you to see what will and won't work when it comes to leading them in the direction you need them to go. What I present here is both the theory and practice and some amusing stories of managing what can only be guided: the Software Development Project.

It used to be that your average business had a single purpose: to sell hardware, style hair, prepare taxes, dry-clean clothes, what have you. Competence in that one single area determined the likelihood of success. If you were a great hair stylist, word would get out and soon you had a waiting list a mile long. Today things are more complicated. Customers now expect a business to have an informative, easy-to-navigate, exciting website.

I recently found a deal on the Internet for a discount on firewood, and then I found myself indignant that an organization that makes its living cutting, splitting, and hauling firewood did not have a decent website. Forget the fact that this operation involves 95% unskilled labor. Forget that there is no real qualitative difference between hardwood delivered from Vendor A or this guy. There I was thinking to myself, "What a bunch of losers! This website is ridiculous! Static pages? Times Roman font?" Logically, the website has no impact on the quality of the firewood that will arrive at my door. In fairness, these guys *had* a website. That should have been enough for me. On the website there's even an online form on which you can order your firewood. There's a page on the ins and outs of proper firewood. It talks about the difference between hardwood and softwood and proper seasoning so it will burn well. It explains how to tell if the wood is properly seasoned. It tells you not to accept unseasoned wood and to make sure to reject it before any comes off the truck. It is an informative site. But its presentation left me questioning the qualifications of the firm to dump firewood in my driveway! Expectations today are high. I wanted to see a function that allowed me to see available delivery times and select one. I wanted to see pictures of good and bad firewood.

In addition, the expectations for technical savvy and capabilities continue to increase. Banks want small businesses to do all of their work with them online via automated systems. They don't want to expend the labor to talk to them. They really don't want them dropping the deposit bag into that slot in the front of the bank at night—they'd cement that over if they could. They want businesses to apply for credit lines, credit card processing services, and additional accounts, but they want them to do that online. They expect that your shop will have Internet connectivity with decent bandwidth.

Insurance companies don't want to talk to you; they want you to go to their website and fill in your business insurance application. The city and county want you to apply for your business licenses, pay your taxes, and file your permit applications on their e-Government site. So the fact of the matter is that even if you run a shop like a firewood delivery service that consists of a large lot with a log splitter, some trucks, and a bunch of burly folks to move firewood around, you're expected to have a fairly complete technical infrastructure. And that involves dealing with engineers—several different kinds of engineers, in fact! And it appears this situation is only going to become more and more common.

Most people find dealing with engineers a trying experience. Engineers seem so difficult. They are distant, hard to reach. They don't make eye contact. They seem to have no understanding of business. They speak so oddly. They make jokes with references to things you've never heard of. They're expensive, which makes it important to get things right as quickly as possible so you can get them off your nickel. You are not alone in this belief. Companies that provide custom software and websites also find working with engineers difficult. Many professional software project managers have turned to drink under the stress of working daily with engineers.

For the past 30 years I've been engaged in the software business. I started out a programmer and drifted to the "dark side" of project management. I'm a bit of a half-breed. To business users, I appear to be a nerd, though not as annoying as most. To engineers, I appear to be a "suit," though not as annoying as most. This no-man's-land existence has offered me insights into the different personality types of these two groups and the resultant disconnects in their interactions.

In this book I offer information, explanation, tips, and techniques that can help managers avoid failed projects and frayed nerves. It is intended for anyone who will be leading engineers in software projects. If you are a small company who has hired a freelance web developer, then you have just put that software project manager hat on your own head, and you would do well to read this through.

Project management is all about manipulating people, which sounds horrible, but let me explain. Manipulation is a term that we most often use in a negative context. That's unfortunate because manipulation can be a positive thing, too. When it's positive, we call it leadership or charisma or team building. We are appealing to the specific motivations of an audience to achieve an end. If your goal is evil, then yes, manipulation is bad. But I'd like to restore the word to its value neutral.

Managing and manipulating engineers is all about knowing what they value, what they like, what they hate, and what they honor. It's also about understanding how they see and experience the world, and where there are differences in those perceptions that might not be obvious to the average person. This book is a step toward explaining that. I liken it to Deborah Tannen's *You Just Don't Understand: Women and Men in Conversation,* or perhaps more like John Gray's *Men Are from Mars, Women Are from Venus: The Classic Guide to Understanding the Opposite Sex.* When you finish this, you should better understand how engineers are different from you, why that's a good thing, and how to use their talents to make your business users happy.

A brief note on stories and names used in this book

I went back and forth on how to preserve the privacy and anonymity of friends and colleagues in the stories I share in this book. At first I thought I'd pick names for the characters that I never once worked with. But I was afraid that I'd forget someone with that name and they'd be offended. Next I thought about making every character name David. My name is David, so I thought that would be okay, but then I realized that David is a very popular name. I've worked with a lot of Davids over the years, and once again, someone might feel I'm over-sharing their experiences in the office. I thought about reducing every name to Engineer A and Engineer B, but frankly that is hard to read and made the stories very stilted.

In the end I decided to use the names Bob and Betty for all engineers. I have worked with many Bobs and at least one Betty that I can recall, but none of the stories are actually about a Bob or a Betty.

In some cases I have changed the names of companies, venues, or timing to preserve the privacy of individuals and my clients. While the names have been changed to protect the innocent, the stories are real and based on my personal experience unless I've indicated otherwise.

Project Startup

Congratulations. You have somehow become responsible for running a project to develop something. Not sure what you did to deserve this, but you should find out what you did, reflect on it, and promise not to do it again. You've been given complete responsibility, total authority, and utter control of it. You have a group of interested, articulate, knowledgeable business users for the system who know what they want and are eager to participate. They've been asking for this change for a long time and will do whatever is required to adapt to a new business model. You have a mandate from senior management to get this done, and you have their complete backing to make it happen. You have a team of excited, experienced, and talented engineers, testers, configuration managers, business analysts, and technical documentation specialists who asked to be on this project. They have worked together before and were very successful. They know the technical environment and have experience in the complete lifecycle you plan to follow. You have an excellent relationship with the organization maintaining the legacy systems and the operations teams as well.

You've been asked to build a system to support a completely new business model, and there is no need to maintain the data from the systems you're going to replace. You have a supportive working relationship with the organization responsible for information security. Their requirements are well known, and they have a set of approved architectures that you can pick from to implement the new system.

Then you wake up and have to deal with the real world. In over 30 years of building systems, this fantasy world, this world that the Project Management Institute (PMI) and others described, has only manifested itself once. It was a project that went so smoothly that most folks thought we were lying about it. Management kept asking me for more detail because they feared I was hiding the facts in my status reports. It just so rarely happens. So this is a book for the real world. It's about how to manage a real-world project with problems, conflicts, budget shortages, time shortages, engineers who can't be allowed near clients, untrained engineers, obstreperous clients, managers who won't manage or make decisions, IT organizations with surprise requirements, security teams that are there to say "no," bureaucrats who refuse to sign things, and resources that get pulled for other projects. In other words, completely normal stuff.

In my mind, the stuff that most people think of as managing—planning, scheduling and resource tracking, budgeting, and reporting—is about 10% of the job. The rest is trying not to hurt anyone, even if they really deserve it. It's working with people. In this environment, it's working with a host of very difficult people, including the following:

Software engineers, who may have chosen the profession because it affords a remarkably good living while minimizing talking to people.

Business users, who have ideas on how things ought to be done that often conflict with the business itself or with every other business user.

Contractor's own management, who have forgotten or mentally rewritten every discussion held during the bidding process.

Customer management, who don't actually do any of the things that the system must do and have a poor opinion of the team they've assigned to support you.

Operations team, who are incentivized to keep the network and computers up, and therefore see you and your "amusing" little changes as nothing but potential sources of future beatings from their own managers.

Security team, who, like the operations team, are incentivized to prevent anything bad from happening to the company's data. The best way to do this is to lock up the data in a dark room. The worst way is to let you create new ways to access them.

There are loads of other constituents:

- Training departments
- Testers
- Process people
- Quality assurance and quality control departments
- The boss' ill-informed trusted advisers

At some point the project manager will be working and negotiating with all of these people. Sometimes you won't know which hat they are wearing, but these people are on every project.

In the following chapters I'll walk you through the lifecycle of a typical project and introduce the various players when they either make their first appearance or have their greatest role. In each chapter I'll try to explain what their roles are, what their motivations are, what their constraints will be, and how best to work with them.

Project Charter

The project charter is the stuff of legends. Like the early land charters King George granted in the New World, the project charter sets out the broad scope and powers of the project manager, granting the project manager complete control over the project and its people. It articulates the responsibilities of the project manager and the sponsors. And it is as often seen as Big Foot. The only projects I have ever run that had a project charter were the ones for which I wrote the project charter. Everyone thought it was cute and quaint; absolutely no one was committed to it. On paper a project charter is a brilliant idea: have senior management lay out what they expect from everyone involved, empower the project manager to get it done, "pin the rose" on the business units to support it, and set some boundaries on the effort. It may be common sense, but it never happens. I did get senior management to sign the charters I wrote, but that didn't mean anyone believed I had the authority and power listed in them. It changed nothing, so although charters could be a great idea, they are also moot. I would not worry too much about getting a charter. It is perhaps one of the most concise embodiments of how management processes ignore how the real world operates.

If you show the charter to an engineer, you are in for a world of hurt. The reason is that the charter, which is signed by everyone in charge, lays out the rules of engagement for the project, its scope, and the like. It specifies the rules by which everyone will play. The engineers believe it word for word, and why shouldn't they? Management signed it, and the first time that reality does not go according to the charter, they will feel angry, betrayed, and upset. Never show a document that can be interpreted as rules to an engineer unless you are certain that it shows how things will actually be done. If the rules aren't followed, you will unsettle their world and become an untrustworthy, random, unfathomable agent of chaos. That is one of the things engineers hate.

Introducing Yourself to the Team

You're new to a group, or you're taking over a project, or for some reason you may not yet know everyone on the team. In many ways this is the hardest and most important meeting you'll have with the team. You need to introduce yourself and establish the culture you want for the project. You have to sell yourself, at a minimum as someone who's not too annoying, and make sure your expectations for proper behavior on the project are communicated.

I had a great notion one time. I had just been given responsibility for a new organization. They were extremely bright, highly educated, highly technical, and quite successful. Enter this guy from outside the group to manage them. So I had the idea to do a presentation on myself and how I viewed the future. I had information on where I was from, where I went to school, my family, the places I'd worked, and what I liked to do in my free time. It was a colossal fail. Holy cow, did I get it wrong! My thinking was that if they knew who I was, they'd feel more comfortable. Nope. My talking about me made them think it was all about me, me, me. They'd already scoured the web and looked me up. They already knew all they thought they needed to know about me.

The only time they perked up was when I described my decision-making habits, that is, I was a Myers-Briggs N and F (see the section

"The Personality Profile of Your Average Engineer" for more on Myers-Briggs and engineers). I told them it would drive them nuts that I make decisions more on my gut than on loads of data. While they may or may not have appreciated the warning, it still worried them. So when you introduce yourself, don't spend a lot of time on you. Odds are they've already read all the available facts about you on the Internet. Instead do a couple of other things:

Explain how you see your role in the organization: Personally, I feel that a project manager's job is to remove obstacles from the team's path and to make some tough calls along the way.

Explain what you expect from them: I expect a professional effort on every task. I expect people to treat one another well, to respect one another's contributions, and to be supportive of one another. It's okay to disagree, but there's never any reason to attack the person with the idea. I expect people to hit their schedules.

Explain your philosophy on the business: I believe we are here to allow our users to do their job better, more accurately, faster, or more completely. We support our users. We are not here for any other reason. Happy users make a happy manager.

Explain how you work: I'm usually in early and leave around six. I'm better at detailed deskwork in the morning and better at people and user problems in the afternoon. If you want to review the budget or schedule, let's do it early in the day.

Explain how you expect them to adapt to how you work: I don't expect everyone to match my hours, but if we have to have an occasional meeting that's either earlier or later than your hours, I expect you to be there. We will try to accommodate your schedule, but if we can't, I expect you to accommodate the work.

Finally, let them know what your absolutes are: I absolutely must have people be honest with me and everyone else. I absolutely must have them record the hours they work and not record hours that weren't worked. I must have respectful behavior among the team members and absolutely with the users.

So What Is an Engineer?

Understanding the nature of the average engineer will go a long way toward understanding how you must behave to work with them effectively. Engineers are not exactly rare, but they also aren't common. If you are not an engineer yourself, you may not know a lot of them because they tend to clump with their own kind. "Birds of a feather" and all. Not many people understand them, so to avoid the stress of a clash of civilizations, they stay by themselves where everyone understands their customs and expected behaviors.

Engineers like to solve problems. They love problems. They love puzzles. One of the reasons they are drawn to computers is that a computer program is really one large logic puzzle, the goal of which is to produce the desired outcome from the user's action while navigating through the data that support the system.

Engineers are usually very bright, and they take their greatest pride from their brightness, knowledge, and cleverness. Whereas fashion models might greatly value their personal appearance or football players their athletic prowess, engineers value their smarts. If engineers had the choice of receiving a disfiguring facial scar or a brain injury that would reduce their IQ to average, they would choose the scar every day of the week.

Engineers like to find the *best* solution to a problem. It's not enough to find a solution that works. They strive to find the optimal solution. They believe in their hearts that there *is* an optimal

> **Glucerna Boxes Are Evil**
> My wife has a Glucerna vanilla vitamin drink every morning for breakfast. It comes in a case of 24. Nothing but the 24 Glucerna bottles packs efficiently into the Glucerna box. Nothing! Bottles, cans, newspapers, plastic containers—nothing. It is therefore inefficient for use as a temporary recycling container, the optimal purpose I put to most cardboard boxes that enter the house. They don't even fold up and nest in other boxes well. Therefore, I hate Glucerna boxes.

solution, and they want to find it before anyone else does. Did I mention that they are competitive? They are very competitive. One source of friction among engineers is agreeing on what exactly is to be optimized. Is it the number of operations required to reach the conclusion, the number of hits on the database, the number of parameters required to be passed, the network traffic, the elegance of the data structure, the flexibility of the service, the amount of code, or any of a thousand other things that could be optimized? This might be the single greatest time sink for a technical manager: making clear what is to be optimized and preventing an engineer's preferred criteria from derailing the project.

Although I am a half-breed, I suffer from this optimization bug in a big way. Every time I go from my living room to my upstairs bedroom I think about what I should carry with me to drop off upstairs and what I should bring back down with me. I think about which rooms I'll pass and whether I should take the extra few steps and drop off an item on my way or check on something. I scan the rooms as I walk, looking for other tasks I could do while I'm walking upstairs to make my walk just that much more productive. I'm really frustrated when I forget to take something up with me and have to make a second trip, not because I hate to walk the stairs but because I did not optimize my journey. Engineers are always thinking this way. You'll hear them debating the optimization weights:

> Bob: "If you went out the west door, you could drop off your letter in the mailbox at the back of the building on your way to get a sandwich."
>
> Betty: "Yeah, but then I'd have to walk another half block to the deli."
>
> Bob: "True, but the pick-up time for the rear mailbox is 10 AM, but the front mailbox is not until 1 PM."
>
> Betty: "Agreed, but it doesn't have to be there until Tuesday, so that's moot."
>
> Martin: "The back door is closer to the CVS, which has a dollar-off special on Pop-Tarts and still had a mess of apple-cinnamon ones when I was there yesterday."
>
> Bob and Betty: "Really!?"
>
> Martin has won the optimization derby.

Engineers might not be vain when it comes to their clothes or personal appearance, but they are extremely proud of their code. They want it to be perfect. Perfect is a tough thing when it comes to code. The problems come from the definition of perfect. To the manager, perfect code does what it is supposed to in an acceptably fast and reliable manner, but to the engineer it might mean something completely different. Once engineers who are any good at all get the code working, they have now learned enough about the problem that they've found a newer and more "perfect" way to do the same thing. If they could redo a section that had something "kludgey" or "clunky" or inelegantly or inflexibly handled, it would have none of that. So they want to go back and perfect it. It's like a model seeing a loose strand of her gorgeous hair and wanting to pop that hair tie off and get it right. Doesn't matter that no one can see that hair when she's walking down the catwalk. To

> **WORDS TO LIVE BY:**
> *Perfect is the enemy of good.*
> — Voltaire

her it's imperfect. (Okay, I know a model probably doesn't have a hair tie in her hair on the catwalk, but I was trying to find an example we could all understand.) Once engineers begin redoing the code, they will find new imperfections that will irk them even more, not just because they are imperfections but also because they did not see them before and therefore the code is not optimal.

Should they just relax a little? Let it go? Stop being such fussbudgets? Well maybe, but they can't—it's who they are and it's one of the reasons they are good at their job. Computer programming is a logic exercise. Computers don't know a thing. They are machines and will do only what they are instructed to do. You have to tell them everything. To a person in the real world, you could say, "Please put that letter on the desk over there." Humans know that you are referring to the letter they are holding. They know that the thing that looks just like a table, a workbench, or any number of other pieces of furniture is a desk. They know that you're waving your hand to the other side of the room and therefore you mean a desk in that general direction. They know that when you said, "put" it, meant that they'd let go of it and have no further contact with it. They infer that you most likely meant the open space on the blotter. They know that you meant the whole letter should be put on the desk, not just the margin.

They know that if it fell off when they were doing it, they should pick it up and put it down properly. They know that if someone walks in front of them on the way to the desk, they should halt, wait for the person to pass, and then go to the desk. They know that if there is no room on the blotter, they should put it in the inbox, or if there isn't an inbox, some other obvious spot, or lacking that, the desk chair. They know what an obvious spot is.

Computers, however, know nothing. They have no idea what an office is, what a desk is, what a letter is, to whom you're referring, what an obvious spot is, what over there means, what "put" means. *Nothing.* The programmer has to explain every single step to the computer. They have to imagine every possible contingency. What if a person walked in front of you? What if the letter falls off the table? What if someone has a desk fan and it keeps blowing the letter off the desk? Check to see if there's a fire burning on the desk, and don't put the letter on it if there is. What if the fire alarm goes off on the way to the desk? What if the desk is moved while en route? What if there's someone sitting at the desk? What if someone made a desk out of water? *Everything.* They have to do this with a pretty rudimentary set of instructions. Imagine explaining how an internal combustion engine works to a four year old. Imagine explaining it clearly enough that the four year old could build the engine. It's doable but you really have to work at it. That's about the level of sophistication that the most advanced programming languages have.

Now add to this complexity the need to do things quickly. Sure computers are quick, but they too can be brought to their knees if the programmer doesn't do things optimally. So having a person who is very detail oriented, craves a puzzle, can think in terms of how to explain things to a four year old (actually a four year old with the inability to remember anything), and will keep hammering at it until it is fast and perfect is a wondrous thing. It's amazing any such person exists. But these skills come at a cost. The cost is that they generally aren't as good at other things.

I've met very few good engineers who were not also very bright. They got excellent grades throughout school, though they might have lagged in writing or art. They were the quiet, bookish sorts. We revere people who are good at math and science, and engineers are good at both. At the same time we resent their abilities and call them names like nerd, geek, egghead, and the like. They remember this. Since the Internet boom, there's been a certain positive patina associated with computer engineers, which has softened the blow of these slurs and even made them into a positive, but I'm getting ahead of myself.

Most engineers grew up doing well academically in school but not so well socially after about the third grade. At that point they noticed that most of the other kids were trying to sap their energy, did not behave in expected patterns, and therefore were to be avoided.

An engineer would like to be physically attractive, but if it's a choice between physical beauty and intelligence, the engineer will choose intelligence every time. The way an engineer can feel comfortable in the world is to be able to do things in it, which requires intelligence. An engineer values cleverness, deep knowledge of arcane things, alacrity with calculations, and the ability to find the "best" way to do something. Designer clothes, a snappy hairstyle, matching accessories, a winning smile? None of these helps you solve hard problems. Well, at least not computer problems, so these skills are discounted. You will occasionally see a well-dressed engineer, but honestly that is rare. Rarer still is the well-dressed male engineer.

Many engineers are excellent musicians. This seems contrary to the desire not to draw attention to oneself, but somehow music is safe. Numerous studies have shown a connection between music and spatiotemporal and logical-analytical functions, both of which are required to develop working systems. The spatiotemporal functions are those used in solving large, abstract problems, for example, software architectures. The logical-analytical side is used for solving complex equations and the nitty gritty of development. I'm not an expert in such relationships, but they seem to go hand in hand. Engineers generally eschew acting and dance. Although dance involves music, it also involves extroverted behavior in front of…people! This is not a good thing to the average engineer.

To an engineer, the world is a frightening place filled with people who conform to no predictable pattern of behavior and over whom they have no control, plus they have no higher authority to whom to appeal such insanity. It's why engineers love games. Games have a fixed set of rules. There are ways to optimize your position based on these rules. People take action during the game in accordance with the rules. If you're playing Risk and you want to move from Iceland to

Eastern Canada, you have to go through Greenland. That's the rule. You can't just decide to invade Eastern Canada directly. That's crazy. It's not allowed. Another great thing about games is that in most games you either lost or had bad things happen to your position because you played it less optimally than your opponent, or it was the roll of the dice. You didn't lose because you weren't coordinated enough or strong enough or quick enough or suave enough. Games are safe.

The Personality Profile of Your Average Engineer

One popular personality classification tool is the Myers-Briggs framework. Myers-Briggs breaks down personality into four facets (I will not attempt to review all the literature on this subject). I do believe that Myers-Briggs can provide a good vocabulary for discussing some of the differences between your typical engineer and your typical manager. From there I can show how interactions can go horribly wrong if one is not sensitive to these differences.

Myers-Briggs breaks down the personality types into four facets. Grossly summarized, they are as follows:

- **Introversion versus extroversion**: How do you get your energy?

- **Sensing versus intuiting**: How do you process the information you gather?

- **Thinking versus feeling**: How do you make decisions?

- **Judging versus perceiving**: How comfortable are you with plans and with changes in plans?

Introversion versus extroversion: In the Myers-Briggs context, these terms don't mean what we commonly think. In this context, it's a matter of where you get your energy. Introverts get their energy from quiet time alone—from the peace and quiet of solitude, from the time to ponder things in a calm setting. They think first and act only after appropriate amounts of thinking. It's not surprising that most engineers are introverts. They have self-selected an occupation in which they spend most of their time thinking of the best way to solve a problem and interacting primarily with a machine. They interact most of the time with the computer, not with people.

Myers-Briggs extroverts, by contrast, get their energy from being with other people—from interacting with people. Extroverts also gain energy from acting on things. If they can't act for a period of time, they actually lose energy. They are frustrated when they can't move forward because someone wants to study a problem "to death." They are gregarious people who want to take on a problem and get it solved. They are willing to go forward with incomplete or uncertain data. They are loud and bold and talkative. They are seen by senior management as brave leaders for these traits.

Introverts are sapped of their energy when around extroverts. Extroverts have no idea they have this effect on introverts. They can't imagine it. They think they're *giving* energy to the introverts by being around them and pumping them up with their talking and bold decisiveness. At the same time, they don't understand why the introverts are denying them energy. The extroverts are working with a whole team of people who won't give them energy. All of this happens at a subconscious level, but it's an underlying cause of tension and stress.

Introverts are hesitant to act without thinking things through thoroughly. This hesitancy saps the extrovert. The extrovert thinks the introvert is overanalyzing things and has no guts to make a decision. On the flip side, the extroverts are perceived by the introverts as fools who go off doing things willy-nilly without reasonable thought.

> E-type: "HA HA! I have relieved you of the burden of worrying about this decision...you did seem to be worrying this to death...I've decided on this bold course of action and TA DA! And you're welcome!"

> I-type (silently to self): "Are you kidding me? He doesn't know half of what he should. We don't know if this is the right answer or not; we aren't even close to making that call! And this clown comes in here and just says do THIS? Are you kidding me?"

It does not take a Rhodes scholar to see how these two groups will clash. In most organizations, the IT staff has a filibuster-proof majority of introverts, and the vast majority of managers are extroverts.

Sensing versus intuiting: Sensing versus intuiting looks at how someone gathers data. Sensors look at data matter of factly and within the context of the environment in which they were gathered. Intuitors, however, look more at the implications of the facts, what they could mean for the future, how they could interact with other things, and the theoretical meaning of the data. Surprisingly, while one might expect engineers to be sensors, they are actually evenly divided between sensors and intuitors. Most research and development groups have a higher percentage of intuitors, whereas software maintenance organizations tend to attract sensors.

Sensing involves processing the data in the current context with the current goals. Intuiting involves looking at the implications of the data, the connections of the information, and how the data might be affected by changes in the future. So without a predominant personality type to point to, you still need to figure out which facet you're working with or there could be trouble, because these two groups sometimes don't get along. They don't understand each other because the intuitors say, "Imagine if this were a hamburger," and the sensors say, "But it's a hockey puck."

If you assign an intuitor a simple "move this column of numbers over one character," he or she might be tempted to make the code configurable so that in the future one could just change the parameters on the fly and have the column show up anywhere you like. The intuitor might think this is a good thing, and that's not wrong, but...it might not be what you want your engineer to be doing. Maybe hard-coding the column over a character is all that's needed, but you have someone spending a whole day making it dynamically configurable.

Likewise, if you ask a sensor to set up a page so it has three main data areas to get feedback from the users, you might wind up with a block of code with fixed widths that must be recoded every time you meet with the users. "The data said three data areas. You got three data areas." Done, but also not what you need, so it's a good idea to see who you've got working and assign tasks appropriately.

Tell the intuitor, "This is a straightforward task that we don't want to spend a lot of time on. It's not going to change." Tell the sensor, "We need three data areas for this mockup, but it's going to change, so don't lock it in." Help them to help you.

Thinking versus feeling: This measure gets at how people make decisions. Thinkers make decisions based on the facts at hand, analytically. They try to leave their emotions out of the decision. They go with the data they've gathered by one of the previously discussed processes and decide accordingly. Feelers tend to use the data as a general guide but make decisions based on their gut feeling, a more emotional approach. Most engineers are thinkers. Many managers are feelers. Our culture is a little confused about how we feel about thinkers and feelers. We love Star Trek's Mister Spock for his reason and logic and use of data to recommend approaches, but our hero is Captain Kirk, who goes with his gut and still saves the day. For this reason we love to have thinkers in actuarial work, in analysis, but we generally choose feelers for our managers and decision makers.

Managers expect their engineers to be logical and offer solutions based on that logic. It's accepted behavior. They believe they are talking to Mr. Spock (but without the ears).

Engineers have a tough time following the decision-making behavior of their managers. It appears random, unfathomable, without basis. They distrust management decisions for this reason. If they don't reach the same conclusion based on the data on hand, they fear their management has made a mistake or is withholding information.

When I introduced myself to a team and explained my strong feeling nature and how it was going to drive them nuts, they paled. It doesn't matter that I had a great track record. To an engineer my track record, given that I have this trait, must have been luck. It's like flipping a coin 50 times and getting heads every time. It's statistically possible, but assuming the next flip is therefore going to be another heads is not something they're going to bet the farm on.

Judging versus perceiving: This measure gets at how comfortable someone is with planning versus "winging it." Judgers gather enough data to make a plan and then stick to the plan. Perceivers keep gathering data and keep changing the plan based on that changing data. The vast majority of engineers are judgers. They like setting the goal and the plan and then using that as the success criteria. Many analysts and managers are perceivers.

You can see from these categories that there are some very strong differences in how your typical engineer sees and operates in the world and how management behaves.

Engineers get their emotional energy from quiet time away from people. They look at data in the context of the problem at hand, and they set a plan based on that data that they want to follow to completion. They want to make a single decision based on analytical processes and see it through based on a fixed set of criteria for success. They are often disappointed.

Management needs to spend time with their people to gain their energy. They tend to make decisions based on their gut, and they keep looking at data and changing the plan.

Looking at the energy, it's no wonder software project managers tend to hang out with other software project managers. The engineers are drained of energy by their presence, and it's the only crowd they can generate energy within without getting a bunch of nasty looks.

From the engineer's perspective, management coming round to chitchat saps their precious energy. Why do they keep doing this? Why are they draining my energy when I have so much work to do? Why do they choose the most difficult and risky occasions to spend all this time with me?

When managers see that their engineers are stressed from working on the toughest tasks, they think that by spending more time together they can rouse the troops and give them some pep and support. When the team responds by becoming more and more hostile or sullen, managers think the solution is to spend *more* time with them and in the process continue to piss them off.

Engineers are fine with the early stages of a project when they are learning more and more about the problems of the business and getting to use their enormous brains to come up with brilliant, clever, efficient ways to solve them (they are thinkers). They love the initial data gathering (they are judgers). The team will usually come up with a plan on how to solve the most important problem, and they rejoice in the mission. "We've set the plan!" So the team is in high spirits. The interaction they might have had with the users was focused on data gathering and therefore was not as draining as most interactions are. They are firing the parts of their brain that they so pride themselves on. They are getting to demonstrate how smart they are, and usually the users are excited that there's all this energy surrounding solving a problem in their area. "Why, that would be great! If you could do that, it would make my work so much easier! Can you even do that?"

This is the emotional acme of the engineer on a project. It tends to go downhill from here, because projects live in the real world, and in the real world things change. Business needs change, budgets change, people just plain change their minds. I've seen users get so excited about an interface design that they went to their boss to show it off, only to tell me a week later that they don't use those functions...ever. It happens. What it means is that projects can fail if the team insists on not deviating from the original goals and plan if needed. They will have to change direction, and, as the manager, you will have to help the team get over the angst those strong judging personalities will generate.

Interacting with Engineers

Now that we have some insight into the mind of an introverted engineer, we can begin to discuss how best to interact with that person. You must start by accepting that it is a waste of time to hope that once engineers get to know you things will change and they'll act "normal." They are precisely who they are, and they are not going to change any more than you are going to become more introverted by hanging around other introverts. Accept the engineer for who he or she is: a talented and gifted individual who sees and interacts with the world differently from you and whose help you need. I begin with what is often the first interaction a project manager has with an engineer, the interview.

Interviewing an Engineer

One of the most stressful moments for an engineer is a job interview.

It's not a lot of fun for anyone, but it's truly hard for an introvert. This is where their worst side might show and they dread it. At some level, almost all introverted engineers know they are not at their best with new people. In an interview, they have to try not just to get by but to shine. They have to show up in their nicest clothes and try to be charming. I have seen engineers in visible pain trying to make eye contact and sit in a relaxed manner in clothes that they won't wear again until their next interview. It's a terrible sight. But there are a lot of things you can do to make the situation better for everyone.

Be Ready

The first thing—and this is true for any interview—is to be prepared. Study the resume before the engineer arrives and know what you want to learn more about before the meeting starts. Some set of items on the resume inspired you to ask the person to come for an interview in the first place. What more do you want to know about those experiences? What technical skills that they listed are you interested in exploring? What ones do you need but didn't see on the resume? Make a list of the questions or topics you want to cover.

Check out the engineer's public web presence to see what sorts of interests are mentioned. Publicly available LinkedIn or Facebook information can tell you a lot about a person. In particular I try to look for things we have in common. Maybe we both grew up in the same city or are fans of the same shows or the same sports teams. Maybe we both used the same archaic software system back in the day. Perhaps the interviewee is a big fan of a particular game. Any of these things will help you help the person relax if you can show that you understand who they are or what they like.

In theory HR has already checked out if they belong to any extreme hate groups and will guide you on how to interview the candidate appropriately. If not, my suggestion would be not to bring it up.

Welcome Them

Second, you have to make your interviewee relax in their worst nightmare of a situation. If you get the typical clammy, dead-fish handshake, do not make a face or shudder. Most engineers don't like physical contact, and they're nervous, which leads to clammy hands. Make sure that you welcome them in a warm way, right away. They are feeling anxious, so their anxiety will get worse if you act creeped out and you'll never learn what their real skills are. Let them know you respect their skills. "Hi Bob, I'm David. I've been looking forward to meeting you ever since I read your resume." Next, make sure they are going to be comfortable throughout the interview. Help them ask for a beverage if they want one. "I was just about to get a bottle of water. Would you like one, too? We have soda and coffee if you'd prefer." Now they are just helping you get what you wanted, and they don't have to put out a request first. You also didn't pin them down to one beverage. I know some engineers who have one and only one beverage that they live on. If I offer them something that is not that beverage, I've actually put them in a corner. Now they have to say that they live on Dr. Pepper and reject my offer of water.

The next step is to bring your candidate to a quiet place with some privacy. No one wants to be interviewed in a big open room. It's unsettling for extroverts and terrifying for introverts. Some people suggest that sitting in an informal, comfortable seating area is the most relaxing thing but not for engineers. They feel exposed and unprotected, and the rules aren't clear. Instead put a table or desk between you to give them some physical space.

Help Them Understand What's Going to Happen

The next thing I do is to lay down the resume they sent and acknowledge it. "I have a hardcopy of your resume here. Sorry, but I made some notes on items that really interested me so I'd be sure to remember to ask you about them." Now they know you're going to be talking about the resume items. It's not going to be small talk and chitchat that they aren't good at. If you're the project manager, you might be the first person that they are speaking to. Let them know who you are and what your role is. It will help them know how far they have to dumb-down their responses. "Bob, I'm the project manager for this effort, and I saw that you had some experiences and skills that we need, which is why I asked the HR department to invite you in." The word "need" is essential here. Some managers are more comfortable with "want" or "looking for" or "interested in," but establishing that Bob is filling a *need* will do far more to calm Bob down than some softer, vague term. It tells him there is an opportunity to be the technical star, to wow you and the team with his techno-mastery. Now help him understand what you want to know. "Bob, I liked that you've done <fill in what you liked> and wanted to learn some more about those projects." Now we've tightened it down even further. You're going to ask them about projects A and D. Great! They know those. They can talk about that. Phew!

Find Out What You Need to Find Out, Technology-wise

I like to help candidates understand what I would be asking them to do on my project when I'm exploring their experience. "Bob, on this project we have a huge data challenge. We have three peta-bytes of legacy data that we have to process into the new system, and we have to do it while keeping the legacy systems running in parallel for up to three months. I see you have a couple of projects in which you used ETL tools. Can you tell me more about the problems you faced with the data there?" If they start talking right away, let them tell you all you ever wanted to know about their past dragon slayings. Try very hard not to interrupt even if they are going off into the weeds. You'll learn a great deal about what was interesting to them and what they think they are good at. If it matches what you need for the project, you're in hog heaven. But you won't get a much better opportunity to find out what they take pride in than by listening to how they describe their solution to a tough problem.

It's great when you have candidates who are comfortable enough to start talking. If they aren't, you have to lead them along. Remember, you aren't hiring them for their ability to sell cars. You're hiring them to solve problems and write code in the process.

Me: "I see you have a couple of projects in which you used ETL tools. Can you tell me more about the problems you faced with the data there?"

Bob: "What do you mean?"

Me: "At Acme when you were using the ETL tool, were there some parts of the transformation that were particularly difficult or that the ETL tool could not handle?"

Bob: "Some."

Me: "Interesting, how did you solve those."

Bob: "I had to write some code to do it."

Me: "No kidding! So what didn't this commercial tool do that you had to do yourself? We were thinking of using that tool on this project but

I don't know a lot about it first hand."

Bob: "..." and at this point you should be good to go. You've told Bob that he is wiser than you are about the tool, which explains to him why you're asking a painfully obvious question.

The point is that to the introvert words come dearly, and you are asking for a lot of them. Help guide the candidate through what you want to know.

Learn about Their Work Style

After I've walked through the technical part of the interview, I then want to explore how this person likes to work. This is tough because a surprising number of engineers are not consciously aware of when they are most productive.

Me: "I was once on a project where we had to work in one large room at a conference table, and we were sitting nearly on top of one another. For me this was completely distracting, and I was begging my boss to let me have a quieter place to work. Did you ever have a situation where you just couldn't get anything done because of how the place was set up?"

Bob: "Yeah, I suppose."

Me: "What kind of situation was that?"

Bob: "Oh, it was just a loud place."

Me: "Do you work better when it's quieter?"

Bob: "Mostly, but it was more the number of people."

Me: "Too many or too few?"

Bob: "Too many."

You've now found out Bob doesn't like large teams. I also like to find out when the candidate is the most productive.

Me: "Most of the team gets in around 9:30 in the morning and stays until around 6:00. Do you do your best work in the early part of the day or later in the day?"

Now unfortunately you have a 50/50 chance of getting an "I don't know" from Bob, and odds are he just doesn't know. If so, then try "Well, if you had your choice, what hours would you prefer to work?"

All of this is moot if your project has fixed, mandatory hours, but even then it might help you understand down the road why Bob is not hitting the targets you hoped for him. He's an evening person, and the client requires him to be there in the morning. Adjust your timelines appropriately.

Check Their Alpha Status

One of the most frustrating things to do is separate the alpha engineers. If you are hiring for a position that is truly a workman role and you bring in an alpha dog, you will be creating a situation that is nothing but trouble and dissatisfaction for everyone. So find out how your candidates have worked in the past and how it made them feel. "Bob, on your past projects was there usually a lead architect or designer?" If Bob replies "Yeah, and he was an idiot. He never listened to a word I said, and the project was a disaster because of it. I should have had his role and he should have done the ETL." Now you've learned a ton. If there was a lead engineer and Bob liked him, then you've learned that Bob can follow as well. If there wasn't but Bob was in charge of his bit and still played nice with the other bits, then you are right where everyone wants to be: with an engineer who can take on whatever role the team needs at the moment.

Acknowledge the Person

At this point in the interview you probably have made up your mind yea or nay on Bob. If it's a yea, then it's time to try to woo him. You've already complemented his technical prowess and shown him where he would fit in; now is the time to let him know that you like all of him. "Bob, I liked your resume so much I checked out your public LinkedIn and Facebook information, and I saw that you're also interested in hiking. My wife and I are hikers. Have you done any interesting hikes recently?" and so on. Frankly, this is my favorite part of the interview. All the tension-filled stuff is behind us, and we're now just getting to know more about each other and share our loathing of hills that nearly kicked our butts or trail markers that vanish into the woods. From an engineer, you're likely to get a ton of useful information about hiking gear, too!

Wrapping It Up

When it's time to end the interview, there's a good chance that you'll be sending Bob off to talk to someone else. Politely close the interview. "Bob, I have really enjoyed our talk. The next person who wants to meet you is our lead engineer. I think you'll like her. You and she have a lot in common." Again, notice the difference in saying "wants to meet you" versus "The next person you'll talk to is…" or "the next interview is." Continue the welcoming. Do not walk Bob to the next interview unless you are darn sure that person is at their desk. There is nothing more awkward than having a great talk with someone—you're excited, they're excited—and then showing up at the next person's desk and finding it empty. I prefer to call the next person and make sure they're available. "Bob, let me make sure that Betty is ready for you. I know she had a very packed day, but she very much wants to talk to you."

When Betty is ready, walk Bob to her desk and do a good handoff. "Betty, let me introduce Bob. Bob and I have had a great chat. He's done some really interesting things with that ETL tool we were looking at. I think you'll be interested in some of his war stories. I also found out he's a hiker. We've done some of the same trails." What you want to do is let Betty know you're feeling positive about Bob and also indicate what areas of expertise Betty might want to explore. You also want to pass on to Bob the notion that we appreciate all of him, not just his technical skills, so that Bob will want to take the job if it's offered. Most important, you don't want Bob to tense up again after you've gotten him relaxed and talking, so you want him to feel some continuity to the interview.

Coaching the Team on Interviewing

To interview Bob properly, your engineering team needs two types of coaching. The first is most properly done by HR, and that is an understanding of what types of questions you can and cannot ask in an interview. Engineers are rule-based and will soak up these guidelines pretty quickly. I've never had a problem with folks breaking these rules, and given that the consequences are obvious (lawsuits that could ruin the company), engineers will follow suit.

The other side of the interviewing process is harder to teach. Some engineers have an inclination to rake a candidate over the technical coals, turning the interview into an oral exam, which isn't a good idea. It creates a dangerous, scary atmosphere for the candidate that very few people would want to work in. There's also a chance Bob knows more than his interviewer, will realize this, get disgusted, and decide he doesn't want this job. The better Bob is, the more other options he's likely to have.

What you need your engineers to do is validate that Bob knows what he is claiming to know. Checking for arcane knowledge of obscure functions in a unique library doesn't help with that. A detailed coding test is usually useful only for entry-level engineers. Asking those classic posers to see how candidates think on their feet is also not helpful. I know one engineer who has categorically stated he will not take a job if asked the "why are manhole covers round" question.

If you have the time, walk your guys through what you're looking for Bob to do and the skills he would need to do that job well. If he needs to be a solid Java developer, then have them ask Bob about his Java experience. Ask him about features of the language that he thinks cause the greatest problems or solve the greatest problems. Ask him things that indicate he knows the proper use of the language. Ask him about IDEs or libraries that this project uses or will use to see if he can prove facility with them. Ask about how he learned Java. It's unlikely that anyone will know the exact technical environment of your project, so you want to find out how Bob learns and how willing he seems to be to learn something new. If your project is going to use nothing but JQuery and you're getting a lecture on why ColdFusion is the best thing since sliced bread, you might not have the most flexible candidate in the world.

I like to role-play with engineers the first time I ask them to interview a candidate. I bill it as making sure they know the HR rules, and although that is important, what I'm really after is coaching the proper tone for the interview.

Betty: "What does SQRT2 return?"

Me: "Betty, that came at me pretty hard. Remember we want to woo this person as much as we want to make sure this is the person to woo. Try this instead: 'We're going to be doing a lot of mathematical transformations on the data. Are you familiar with the math library? For example, do you know what the SQRT2 should return?' See, now he knows why you're asking, and it's not just something out of the blue."

You also have to coach them on the transition to the next interview. Help them learn how to be good hosts to the candidate. It's harder for an introvert to be warm and gushy. It takes practice. Give them that practice. For new teams, I've actually been on hand to walk the candidate to each interview to watch and then coach the engineer on how to make that transition better.

Some Common Quirks of Engineers

There are a couple of quirks that engineers seem more likely to have than the average population. There are two that you will likely encounter and should know at least a little bit about.

Asperger Syndrome

Asperger's is in the autism family of syndromes. When people think of autism, they think of Dustin Hoffman in *Rain Man*, but there are many levels of autism. Asperger's is one of the more functional flavors. It can manifest in many ways, but most often it involves awkward social interactions, repetitive behaviors, and odd use of language. People with Asperger's are often unaware of how their behavior affects those around them. They don't have a strong sense of empathy with their audience. They don't read body language, pick up on tone, or see the mood of the person via eye contact. This is different from being shy; it is a true void in their understanding and ability to pick up on these common signals. They can irritate someone and have absolutely no idea the conversation is going poorly, even when the other party glares or snarls at them.

> **Pedantic Word Choices**
>
> We (in the United States) often use "peruse" to mean that we are quickly skimming over something. Someone with Asperger's might be compelled to correct you that its actual meaning is to pore over in great detail, as in an audit. Fun.

Another "feature" of Asperger's can include repetitive behaviors. People with Asperger's might wring their hands continually during a meeting and have no real awareness they've been doing it. It can be distracting to say the least. Finally, they may use odd forms of speech or be particularly precise in their use of language. They might never use a contraction. They may use only the precise primary meaning of a word, even if it's common practice to use the word in other senses. American English is particularly fond of contractions and shortcuts.

People with Asperger's are very likely to correct any use of a word that they don't approve of, even if it has nothing to contribute to the topic at hand. Even though they're right, it can make having a "normal" conversation more difficult.

My point in bringing this up is not to school you on how to be a professional therapist for someone with Asperger's. My nonscientific, anecdotal data would predict that roughly one in fifty of the engineers you meet might have some level of Asperger's, so you need to keep it in the back of your mind when working with someone who is being difficult. They might have some real issues that all your coaching will not solve. You should be aware that their behavior may not be willful; it's possible they can't help it. If your star developer is someone with Asperger's, you may need to keep this employee away from customers with thin skins. It's a tough call, but there may not be a lot of other options. Odds are the other engineers will value the contributions made and put up with the lack of social niceties.

I've had several engineers on my teams with Asperger's. One in particular—I'll call him Bob—was a genius in the truest sense of the word. He could make things happen on a computer that no one else I've ever met could do. If I needed someone to connect my computer to a standard toaster and send the results to my refrigerator, Bob could make it happen. He did things on his own time over a weekend that the whole rest of the team could not have done in months. Bob was that good. He lived for a challenge and he never failed. He also loved the customer, but it took a long time in very controlled meetings for the customer to warm up to him. As a manager, you have to prepare your customer for someone like Bob. You have to explain that he's a little different and share that it was Bob who did the amazing things that the customer loves or is about to love. Let the client know that Bob may be a little different but he's essential to the task. Even the least technical client appreciates hard work, ingenuity, and creativity and probably has worked with someone who is a little different.

You have to control the conversation more than usual. Back in the office, you can let someone like Bob go off on a tear when he needs to. The other engineers know that it's just Bob's way and he gets like that sometimes, but the customer is probably not prepared for Bob. Bob is also more likely to behave quirkily with the customer, because he's stressed by the unfamiliar environment. In the field you'll have to shut him down a lot sooner than back in the safety of the office. You need to pull him aside beforehand and explain that you'll be doing that. "Bob, tomorrow we're meeting with the customer, and they are not very technical at all, much less up to your standard. I need you to contribute but be very careful what you say. We don't want to scare them off. I'll be stepping in and controlling a lot more than usual what everyone says. Don't take it personally. I need you there, but I also need the conversation to go a certain way. Are you okay with that?"

Generally the Bobs of the world will be; in fact, knowing that someone will bail them out of any awkward moments can be a source of relief. I had clients who, after a couple of years of working with software Bob had built, would make a special effort to stop by his desk whenever they were in the office just to have a quick chat. Bob, of course, probably didn't even recognize these people.

Talent, once properly presented, is always respected by a smart client.

Remedial Small Talk and Other Human Skills Coaching

No one would think twice about sitting down with an employee and showing her how to read a Gantt chart or fill in a Risk Matrix. We send people to classes all the time when we switch languages, database managers, or operating systems. It's perfectly okay to do that. People even plan for it in budgets; that's why there's a training side to the human resources department. Now imagine holding a class on being civil in a meeting. What would you call it? "How Not to Be Rude" or "Social Graces 101"? It's not something that I expect to see in any organization anytime soon, but it's actually needed. For some engineers the one thing holding them back from advancing is social graces, or rather, the lack thereof. But this training ought to be offered, because it's needed. My wife did a project at a company that offered an online course called "Working with Difficult People." She insisted that everyone on her team take the course, arguing that "At some point in this project every one of us is going to be difficult people, and we should all learn how to respond and recognize it in ourselves." It did not change the world, but it did help.

If you've managed developers for any length of time, you've likely had people pull you aside and say warm and welcoming things like, "Don't ever bring that guy to our meetings again." Maybe they were more delicate and said something like, "I think the users would be more forthcoming if we kept the meeting less technical. Maybe next time we could just have you and the users come." More often it's blunt: "Bob really pissed off our team. Does he have to be there?"

Well, the fact is that it's always best if the developers can hear in person about the problems and the business that the users need the system to solve. Reading about it on paper, especially in a context-sensitive language like English, can lead to well-intended but unfortunate misunderstandings. In some cases I have had to ban certain engineers from answering the phone if it's a user calling or from going to meetings with users. I was once at a location where users frequently would come to our area unannounced. In some ways it was great, but I had to leap from my desk to screen visitors from one of my engineers. He didn't work at being a jerk; it was a natural gift, really. He was a jerk to us too, but we were used to it and accepted it as part of a package that included some amazing technical wizardry. His behavior wasn't personal. He just couldn't filter anything that came into his mind. He didn't even understand that it might be something one would want to do.

So it falls to you—the technical manager—to coach your people in the arcane knowledge of human interaction if you want to help them advance in their careers. No one will help you out; you're on your own. If you keep in mind the place your audience is coming from, you can make some good progress. Let me begin with the tale of the most extreme example of "house breaking" that I've ever had to do.

As fate would have it, this story is about a woman. We had a brand new employee. She had an excellent resume and some great experience in a number of places. Her husband was in the armed services, and she had to relocate every two years, which explained a nearly-every-two-year job change. Her references raved about her technical prowess. What everyone failed to mention was that she was marginally functional on an interpersonal level. She interviewed in a very professional suit, and that was the last time we saw her in one. She came to work in corduroys and a flannel shirt in an era when everyone wore a suit and tie to work, including programmers. She would enter the building, go to her office, and work all day long, appearing only occasionally to pick up a printout or use the bathroom. Pounding down the hallways, she had a severe scowl on her face. A long conversation with Betty was anything longer than a grunt. Her technical work was superb; I had no complaints about that at all. Her code came out ahead of schedule and was of very high quality. It was rare we found a defect in Betty's code.

Coding was why we hired her, but she didn't get along with some of the folks in the office. That would be the living folks in the office. Every one of them. It was a problem. I was told to handle it, so I started talking to Betty on a regular basis. As I was her manager, she allowed this, though barely tolerated it.

One day we were talking about her attire, which, as I mentioned, was significantly below the current standard of professionalism. This was the era of the three-piece suit. Imagine a whole pack of nerds in snappy vests, polished shoes, and natty ties. It was a scary era. Women (not that we had loads of female engineers) wore skirts or dresses or, as was popular then, masculine-looking business suits. Swell, we have a tiny population of women, and now they're all dressing like guys. But I digress.

My brilliant developer was coming into the office in flannel shirts and corduroys, and although I didn't much care one way or the other, my managers told me I had to get her in line with policy. We did have an official dress code; it was in our employee manual, so I had something to back me up. Don't try enforcing a policy with an engineer if you don't, because until it's a rule, it's not a rule they have to follow.

I explained the situation: that she was not compliant with the employee manual, that it was not okay, and that I needed her to comply with our policy. It didn't start out well.

> Betty: "I don't see the customer. Who cares?"

> Me: "Well, it's not about seeing the customer. We have a policy that we all have to dress in a professional manner."

> Betty: "I'm more comfortable this way. When I'm more comfortable, I'm more productive. Do you want me to be less productive?"

> Me: "No, I'm really happy with how productive you are.

You're amazing, perhaps our fastest developer. Even more so in that your work is extremely solid. We hardly ever find problems in your code, ever. But if you look around, everyone else is figuring out how to produce in business attire. I need you to follow the company policy."

Betty (extremely upset now): "I can't afford to buy all new clothes! If you're telling me that it's either pay my electric bill or comply with this stupid rule, then let me tell you…"

Well, I thought I had some flexibility here and was going back and forth with her about phasing in a more professional wardrobe over time. But I had a hunch that the issue wasn't complying with a policy. That's a rule and engineers *love* rules. There had to be something deeper. She was pushing back on timelines and costs and arguments that did not hold to logic. And I had a hunch…I said, "But Betty, if you just picked up a couple of pairs of nice slacks and a handful of collared blouses, you'd be there." She stopped cold. And suddenly the wall fell down. She breathed again. The really scary part was wearing skirts and dresses. Engineers already don't like to be the center of attention, and in our society it's the women who are the brightly colored birds. Men wear boring, dark-colored suits, and women are pretty and shiny and alluring. Now imagine an introverted engineer being asked to fit that mold and wear clothing designed to attract attention. What better way to induce a panic attack?

I tell you that story as a preamble to our next story about Betty. Working through this clothing crisis established a level of trust between us. I was still a management weenie, but at least I was one who seemed to understand her and wasn't overly annoying. Betty was not unaware of her discomfort with other humans, but she had no idea how to change anything. The discomfort was constant. There was never a moment when being around other people wasn't awkward, scary, or stressful, so there was no one thing she could point to that set her off. In the lab, mice only get shocked when they push on the wrong lever. Inside that brilliant developer, she was getting shocked all day long by everything she did if there was another human in the room.

Remedial Small Talk Story

As I mentioned before, with great power comes great…other things.

Betty was perhaps one of the least sociable persons I've ever met. It's not that she was mean, but she had a perpetual scowl on her face that chased people away. This was a good thing since she found people both annoying and largely irrelevant. Not maliciously so, she just had no need for people. She was also perhaps the most introverted person I've known. By that I mean that interactions with other people were draining…deeply draining. Talking with people left her feeling like she needed a nap. It sapped her blood sugar down to nothing.

One day we were talking about an assignment (an interaction that is tiring but not draining for her), and she complained that everyone avoided her. Our conversation was interesting.

> Me: "Well, I'm not sure you realize this, but when you walk through the halls you often have a scowl on your face. You look angry. It scares people off. People are afraid of angry people."
>
> Betty: "I'm not angry."
>
> Me: "I know that but they don't know that. Everything they see tells them that you're angry. My guess is that if you didn't present them with an angry face, they might not think you're angry."
>
> Betty: "So you're telling me I have to pay all my attention to how my face looks every time I walk down the hall or people will think I'm angry?"
>
> Me: "No, but when you see someone coming down the hall, you could try smiling for a few seconds and see if that changes people's reactions."
>
> Betty: "What would I be smiling at? There's nothing funny."
>
> Me: "True, but again, you just want them to not be afraid you're angry."
>
> Betty: "But I'm not angry."
>
> Me: "I know, but you look angry, so if you try to not look angry when people are facing you, then maybe their impressions of your mood would match your mood."
>
> Betty: "I don't know what expression would match my mood!"

Me: "I get that, but right now you're expression says angry and that's *not* what you are, so we're starting with a mismatch. Any change would be better."

Betty: "Well, I'm not going to walk around with a village idiot grin on my face."

Me: "No, that would be bad. But anything other than angry is what we're after. No?"

Betty: "Okay, I'll try."

So what was I doing? In each exchange notice that I agreed with the logic of her statement (deny logic at your peril) and then went on to describe the world she could not see and focused on the cause and effect, inputs and outputs. You look like you're mad so people think you're mad. She scowled, which made people believe she was angry. What other set of data do they have? None, so they reasonably believe you are angry. Cause and effect.

Do I have to expend all this energy all the time? No, only when you wish to affect the environment you're in, when people are around that you don't want to scare off. I was focusing her on how her actions affected her world and how she could change her actions to effect the change she wanted from the world. I was explaining the system operating behavior of human interaction. I'm not for a second suggesting that the world behaves safely and consistently like a computer does. What I'm pointing out is that you have to relate the "scary" real world to the behavior of something they do trust, which for her was a computer.

And try she did. She tried hard, and it actually worked…mostly. At first it looked like she was having intestinal distress when she walked down the hall, but over time she found an expression that neither solicited needless stress-inducing conversation nor scared people. She relaxed more in the office. She was clearly curious about people, and, as I've said before, introverts want to be acknowledged by other humans; they just don't want them sapping their energy. She was now no longer being shunned, but she wanted that acknowledgement. We were ready for the next step.

Betty paused one day when we were talking about a task and said, "People say hi to me now but then they look at me weird."

Me: "Really? What did you say to them?"

Betty: "I didn't say anything. I don't work with them."

Me: "Ah, okay I think I know what's happening. You see, when people say 'Hi,' they expect to get a 'Hi' in response."

Betty: "Why?"

Me: "Actually, it sounds silly now that I think of it, but it's like a protocol between two people. One person sends a 'Hi' and the other responds with 'Hi.'"

Betty: "But they do it all day long."

Me: "Yeah, you're right. It's how we fill those awkward moments when we're walking toward someone and have no idea what to say, because really, we don't have anything to say, so we say this greeting and we all agree it's less awkward this way....It's like the system idle process!"

Betty: "Oh, okay."

Me: "So let's try it, just you and me. Pretend we're walking down the hall toward each other. I say 'Hi.'"

Betty: "Hi."

Me: "Great! See, it's simple. We filled the awkwardness with a standard protocol!"

Betty: "So what happens if I see them way down the hall?" (This is brilliant! Notice that a good engineer is always thinking of the boundary conditions, exceptions, and bad data.)

Me: "Oh, excellent point. Usually you don't say 'Hi' the second you see someone. There's actually a sort of commonly accepted distance. I have no idea what that distance is in any set of units. It's a feeling. Let's go into the hall and I'll show you what I mean."

And so we did. We went into the hall, and I walked a short distance down the hall and turned around and started walking toward her until I felt like a "Hi" was in order. As an E-personality type, I'm very comfortable talking to people and just had an instinctive sense of when one should say "Hi." Too far back and you then have a long way to be looking at each other with nothing to say. Too close and it feels like an ambush. Imagine walking down a hall until you were immediately opposite them and then saying "Hi." The person would jump. They'd wonder what was up with you. What do you want? No, don't take my wallet! However, to some engineers, this is *not* intuitive. Seeing someone can be draining all by itself. They're thinking "another remora to steal my precious energy." It's awkward from the very moment they spot the approaching person. So this little drill was actually pretty helpful.

We practiced for a while—we really did. Just starting at opposite ends of a long hallway, practiced not scowling and when a reasonable distance was reached said "Hi." She got it right away. What I didn't realize is that there's also a velocity dimension to it. If one person is cruising down the hall and the other is sauntering, you expand your "say Hi" tripwire point. I discovered this on about our tenth practice "Hi." I was trying hard not to lose my marbles, so I was mixing it up by rushing down the hall. I initiated the "Hi" but was then told that I Hi'd way too early. How did she know? Well, she'd watched me and counted how many doorways away I was when I said "Hi." Metrics! Units! Things that are predictable. Things an engineer so wishes we used in human interaction! So I am not anywhere close to the proper doorway and I blurt out a "Hi." Wrong! I had to pause and think about it and explain that there was also a timing aspect of the Hallway Hi. It had to be soon enough so you didn't feel ambushed but not so soon that you would feel awkward afterward. This is a theoretical concept for an engineer given that they feel awkward from the get-go, but she granted that it was possible to have a human who did not, so she gave it to me.

At first I was building up her confidence by showing her that the Hallway Hi was a simple function that could be learned, and then I crushed her by introducing the velocity term to the equation. I had no idea how complicated describing the equations of the Hallway Hi model was going to be.

So now she had the Hallway Hi down, and she wasn't scowling, but she was still a ways off from getting the human acknowledgment she so desired and certainly deserved. "Hi" is necessary for this but not sufficient. We're going to need a couple more lines. I knew we were on the right track when a conversation started like this:

Betty: "I say 'Hi' and then people keep asking me how I am. Do I look sick?"

Me: "Oh no, you look fine. It's just a common thing to do. Let's try it." Based on our previous practice work, I thought she might go for this. I said "Hi."

Betty: "Hi."

Me: "How are you?"

Betty: "Fine."

Me: "Ok, now ask me how I am."

Betty: "I don't care how you are, you look fine."

Me: "Well thanks, but see, it's an interaction protocol. No one actually, or at least very rarely, cares how you are, but it's how we greet each other. It lets us ACK each other safely. We all know what's expected and it's safe. Oh, and another thing, we usually thank the person for sending us the standard protocol. Like: 'I'm fine, thank you. How are you?'"

Betty: "But if it's expected, why do you thank them?"

Me: "That's a good point. I guess it's one of those completely arbitrary parts of the protocol. But it is expected and appreciated. It makes the receiver happier."

Betty: "Well, I don't care if they're happy or not."

Me: "I understand, but you certainly don't want to make them unhappy. You're not a mean person. If you could avoid it so easily by just saying 'thank you,' even if it's not needed, you'd do that, wouldn't you?"

Betty: "I see."

Betty and I spent an hour every Tuesday and Thursday early in the morning before anyone else was in the office role-playing small talk. We talked about how some subjects were difficult and dangerous and others were safe and actually invited the person to share something personal. "Do you have any pets?" Good. "Are you eating more than usual?" Bad. When I hit her with the pets question, she then went on a tear about her pet—a dog—and its breed and the history of the breed, and on and on. We talked about letting others stay connected. How you want to ration out your knowledge to keep the contact going and let the other person participate. Otherwise you've just suckered them into a lecture they didn't know they signed up for. How magic the "Oh! You have a dog? I have a dog too!" is. Now you've shared a connection. Let them tell you about their dog. Then you tell them about your dog. You've both now paid in.

We did this for several months, and she really worked hard at it. At times it was difficult, a little like watching a stroke victim learn to walk again. But to her credit, she hung in there. Her early forays into the hallway were stilted, struggling experiments. She seemed to be a little outside of herself, gathering the data from each interaction and processing it, but over time she realized that most of the folks in the hall also loved dogs, Star Wars, science, and nerdy things. That they were safe and wanted to learn what she had to teach. That they, like her, saw value in the currency of knowledge, and she had a lot, so she was valuable. She got her acknowledgment and loved them for it.

The point of this story is not that there are broken people out there. There may be some, but Betty wasn't one of them. She was uneducated in the ways of common human interaction. She did not have an intuitive sense of how her behavior affected or could control her environment. How could she get through school, college, and multiple previous jobs without better human skills? I can't imagine and it's not important. The point is that as a manager, you have the moral obligation to help each team member integrate into the team. Most of the time you're teaching them how to read the schedule or how to conduct a code review or some other technical skill; other times you're working on people skills. It's all the same.

For those of you paying attention, yes, I think Betty had Asperger's.

Excessive Distrust and the Tale of Direct Deposit

Bob was a great developer but likely another Asperger's case, albeit mild. Bob read every piece of paper he was ever asked to sign, top to bottom. He read every word. If it referenced some other document, you and I would have just said "yeah yeah yeah, whatever" but not

Bob. Bob would demand to see it. I'd have to go dig it up. Ever try to find, read, and then explain the New York State Wage Theft Prevention Act on short notice? Well, neither had I, but there it was on an agreement I needed Bob to sign, so off I went.

One thing that I never knew before I worked with Bob was that almost all direct deposit agreement forms have a clause in them that says if the company needs to make an adjustment, say, if they make an error in the deductions, they can "direct withdrawal" that amount. Most of us would agree that this policy is reasonable, albeit annoying. But to Bob, this was opening up his entire financial security to the random feeling-based decision-making process of corporate suits. It was unacceptable, so he refused. No big deal, right? An employee wants a physical check mailed to him. They probably have a slew of vendors who get a physical check, so there's already a process for this. Again, why bother?

In this case there was a problem. You see, Bob lived a very Spartan life. He had a one-bedroom apartment, no car, no student loans. He did not bet on the ponies or have a bevy of expensive women to entertain. He wasn't into designer clothing (surprise surprise). He didn't have a fancy stereo. He lived for and loved his computers, and that's what he did every day. He built his own machine so he could incrementally update it as he desired for next to nothing. He left work, slinging code all day, to continue such activities on his own at night. There is nothing wrong with that. He had a phone bill and nearly nothing else. So he didn't need a lot of money but was paid market wages and that reduced the necessity to bring those physical checks down to the bank.

He often had a slew of paychecks sitting at home. By a slew I mean that at one time he had 43 uncashed paychecks. We got paid every other week. Yes, a year and a half of checks, uncashed. This created a problem for the accounting department when it came time to close the books for the fiscal year. We'd beg and plead with him to cash the checks. I'd offer to drive him to the bank at lunchtime. Sometimes he'd take a pile and go to the bank, but a month or two later, I'd get the call from accounting asking for help. This went on for over a decade. We went through multiple managers in accounting over that time, and each time I'd have to explain it to them. Each time they'd stare at me in disbelief.

So how did this finally get resolved? It was a logical argument supported by data and a bit of knowledge Bob didn't have.

We explained that although it was true that the company could deduct money from his account for errors they made, they had been making those same adjustments on the physical paychecks all this time. There was no difference. I asked him to check the paystubs from the period when our tax rates changed but the company was a week late making the change, and sure enough, the second check was doubly low. Winner! No. He had a rebuttal argument. The FDIC only insures up to $100,000 per bank account, and, as you can imagine, he would have more than $100,000 sitting in it if he cashed a year and a half of paychecks. But...Bob didn't know you could have as many accounts as you wanted to. We suggested he put some of it in CDs, some in a savings account, still more in money market accounts. He actually perked up! He was suddenly an investment baron. He agreed to a direct deposit into his old checking account. From there he'd quickly move almost all of it to his other accounts and voila! Protection from the evil corporate excesses.

Bob was stuck on the literal interpretation of the direct deposit agreement. Combine that with an overabundance of distrust of managers (probably because they were not predictable and their decisions didn't seem to be based on reason). Bob was finally managing his money and earning something on it. I stopped getting the harassing calls from accounting, and everyone was happy. In fact, accounting took us to lunch.

Hoarding

Until recently the concept of hoarding was rarely spoken of. Today we have several cable television shows revealing how people can be buried under their own stuff. I've seen engineers who "collected" soda cans, newspapers, computer printouts, old nine-track tape write rings, broken disk drives, and disk drive magnets. The specific object of collection is unimportant. It usually isn't a problem unless, as sometimes happens, it becomes a fire or health hazard (pizza and fast food boxes) or impedes the employee from actually using his or her desk. Yes, it can get that bad. I am not a psychologist, and I won't try to explain the underlying reasons for hoarding, but I will say that you should not over-react when you stumble upon it. It's out there.

However, I would recommend handling it gently, with respect and care. There is a "need" for the employee to have all those newspapers. I don't understand it, but it's real to the employee.

> Manager: "Bob, the stacks of newspapers you have here are impressive. You've clearly been collecting these for a while. Unfortunately, I need you to move these newspapers out of the office. We've been told they're a fire hazard."

> Bob: "What! They're not a fire hazard! We don't have any open flames or machinery in here!"

> Manager: "Well, they are made out of paper. The fire marshal wasn't happy about it. It might take a while to move them out, but as long as we get going on it today and can show that we're not ignoring them, I think they'll be okay with it. If you keep it to one or two days' worth of papers, I don't think they'll bother you in the future."

The key here is that you don't want to say anything like "You need to get rid of these papers!" or "You have to chuck this crap out now!" It's not crap to Bob. For whatever reason, it's a treasure. Ask Bob to move it. If you can reduce it to "This is a rule. I know it doesn't really make sense, but it's a rule," that helps get the rule-followers to accept the request rather than argue with you about whether the request makes sense. You can shortcut a lot of arguments if you agree that something doesn't make sense based on the data (e.g., we don't have open flames here). But "The fire marshal has a rule about stacks of paper" is something that doesn't have to make sense for the engineer to comply.

Do you care where? I hope not. What you need is for the papers to leave the building. If your engineer takes them to his home, that's his business. I have had someone take them to an empty office, and boy, was that a fun chat!

Pace Workers versus Burst Workers

Engineers, like all humans, work in different ways. When I have a task to accomplish, I tend to think about it, ponder it, and do a lot of things that have absolutely nothing to do with it until my muse gets my attention and I sit down and bang it out in no time. I'm a burst worker. To an outside observer, I'm goofing off most of the time and only productive for 10% of the time. Now the reason I can just sit down and bang it out is not that I'm generally brilliant but lazy. It's because the production of it happens well after I've thought the problem through. For each chapter in this book I mulled over what I wanted to say, what points I thought were good. Over at least a week or so I would think about how to phrase something and edit it in my head. I'd go through several approaches that all led to dead ends before I ever sat down in front of my keyboard. Lots of this pondering happened while I was doing something completely unrelated. Then I'd bang it all out and voila! This way of working drives project managers completely bonkers. They want their team sitting at their desks churning out code at a predictable rate. How else can you know whether you're on schedule?

Project managers are much more comfortable with pace workers. A pace worker is precisely what project managers think they want—people who sit at their desk and code all day long. They never get up; they aren't chatting away or "goofing off." They are sitting there working. The irony is that my experience has been that neither type is more productive than the other. The burst workers I've supervised crank out exactly the same amount of work as the pace workers. They just get it done differently.

As the project manager, you need to recognize which employee is which and handle them differently. For your pace workers, you need to remove distractions and interruptions. They need every minute of their desk time to be focused on their work. They have a general production rate, and you can learn what that is over time and use it for planning purposes. Managing them is easy but can be frustrating if you try to compare them with a burst worker. When burst workers are bursting, they are cranking out all manner of code very quickly, but on a daily basis you see guys at their desks all day who are just not as productive…to the untrained eye.

Burst workers need to let their subconscious work the problem. You should assign burst workers some distractions to allow time for this background processing to work its magic. An architecture diagram for a security review is a great task. Bullet points on some of the risks you're working. A review of some other documentation can facilitate this. The other thing you have to do is trust them to be working when all outward appearances suggest that they are not. Let them talk to the testers about the upcoming soccer matches. Don't worry when they're chatting about the weather in your office. Trust their process and let them do what they need to do. One thing you do have to be mindful of is not to let your burst workers distract your pace workers. You'll annoy the heck out of the pace workers with interruptions. At the same time, the pace workers will want to talk about the work and this will undo the background processing for the burst worker. Both lose.

There is a risk here that your pace workers might perceive your chatting with the burst workers as favoritism. They might think that they are the only ones working while you and the burst workers are yucking it up all day. As a burst worker, I've always found it helpful to announce myself as one. I acknowledge the existence of such a beast, and it tends to go a long way toward calming these sorts of frustrations. If your burst worker is not hitting milestones, then you might not have a burst worker at all. You might just have a chatterbox, but that's another story. If everyone is hitting deadlines, then life is good, and this is the culture you need to instill in the team.

Never Lie to an Engineer

Engineers like things to be ordered, understandable, and predictable. They take great pride in their understanding of cause and effect. They are rule-based. They love simple systems in which you push down here and it goes up there. They live for the challenge of a complex system in which the outcome of any given input is known, but you have to be very clever to describe how that happens. They work in a world where the data and the rules really matter. If an engineer makes decisions based on false data, a bridge may fall down. Engineers have to be able to trust that their inputs are accurate or there are real consequences.

They hate chaotic systems. Chaotic systems are those in which any given input can produce any random output; one in which cause and effect are decoupled. It upsets their world because they have no control.

Management often makes the mistake of thinking the engineering team is ignorant of the workings of the business, and frankly sometimes they are. But they were hired because they are clever and can solve problems easily. So if management gets up at an all-hands meeting and tries to "sell" a cut in benefits as a good thing, they usually just piss off the engineers. Management is not being honest. They are not calling the situation as it really is.

It is absolutely okay to tell a group of engineers that they are going to get a cut in benefits because health insurance premiums have gone up 13%, so that's the best deal management could negotiate with the insurance company. Cause and effect. They won't like it, but they won't think any worse of management.

However, if management gets up and tries to put lipstick on that pig, they are in for trouble. "We've got some great news! We have a new wellness program in which we have employee-centric benefits. Now you will control where your benefit dollars are spent. Your new premiums will be…" and it's truly the same cut. The engineers started running the numbers the minute you started talking, and as soon as that slide went up they figured it out. Management is lying, which is the worst thing you can do to an engineer.

You are now a chaotic system. Nothing you say can be trusted…about anything. Sports, the weather—you name it. Lie about one thing, and—to an engineer—all your outputs are untrustworthy. Furthermore, since you have so unsettled their world, you will likely never be trusted again. It's akin to hiring a babysitter and finding when you come home that she's been having a party at your house, your best booze is gone, the couch is on fire, and the kids are cowering in the corner. Will you ever hire that babysitter again? That is how a lie is perceived by an engineer. Don't do it.

I had a rather spirited discussion with the managers of a startup company for which I led the technical team. The coffers were empty, we were six weeks from delivering our first system, and we were out of money, or nearly so. Our vulture capital investors were extorting us for the next round of financing, and they knew we were between a rock and a hard place. If we yielded, then everyone's stock would be nearly worthless. Some didn't want to kick in the next round at all. So it was very unlikely we were going to make payroll that Friday. I argued that the team was committed to the project, that they wanted this to be a success, and that if we told them the truth they'd be unhappy but would support us. Management didn't believe it and insisted we tell them nothing.

Now based on the engineers' behavior I described earlier, the best situation we could possibly have is that everyone's feelings about management are the same, but now the company payroll system becomes in their eyes chaotic, that is, bad.

But no, management dug the hole that much deeper by holding a meeting and telling everyone that they were doing great work and that everything looked great for the future. They lied. On Friday there was no money to pay them. The negotiations with the vulture capitalists were going better, but there was nothing solid on the table. So they broke out an excavator and dug a huge hole they could never get out of. They told the team that although there was no check today, they felt sure that by Wednesday it would be okay. They lied.

So what did I do during all of this? I told them facts, not lies. I told them what I thought was fair to share, but I did not spin it. When Friday rolled around and my team didn't get their checks, they came storming in to me. I told them that I too didn't get paid. I told them that the vulture capitalists were trying to reduce our ownership to nearly nothing. I also told them that we didn't have a lot of leverage given that the product was not out and we didn't have any revenue streams, that the vultures knew this and were trying to take us. I told them that I thought we'd come to some agreement because if everything stopped today, no one would make a penny. Both sides knew we had to get the product out the door. I told them that if the talks completely broke down and we could not find any new investors, we'd have to liquidate everything, and since the only thing we had of any value aside from some well-worn PCs was the product, we should get that out as soon as possible no matter what. These were unpleasant facts but true nonetheless.

The team never forgave management. They never trusted anything they said again. Whereas before they spoke with management about as much as any engineer might (i.e., not much), they now refused to attend meetings if management were there. If forced to attend, they would say nothing.

Tell the team what facts you have. When management was making cuts in overhead budgets that were going to reduce my team, I told them revenues were down, so our overhead had to be reduced. We didn't know yet by how much, but from what I knew at that time about our expected revenues, it was somewhere between 10% and

20%. Based on our current staffing profile, that would be between 2 and 4 overhead slots that would have to go on direct billing or, if there's no direct work, to another division, or, god forbid, out the door. So the very next question was, "Yes, *who*, is going out the door?" I did not know, so what did I say? "I do not know and we don't yet know about direct work, the exact percentage, or if our group will be hit as hard as the others. I know this is upsetting but before we panic, let's find out how big a problem we really have." And perhaps the most important thing you can say to an engineer: "I will let you know whatever I find out, as soon as I find out." But you have to mean it.

It's even better to be honest if you've been forbidden to talk about it. Engineers can respect that there are rules. Don't lie about the situation if you've been told not to pass the facts on to others. It's so much better to say, "I've been told I can't talk about it yet. I don't know when they will let me talk about it, but as soon as I get the okay, I will share what I'm allowed to share." Better that than "Things are fine. There's nothing to worry about."

So never lie to an engineer.

Helping Engineers Through the Learning Curve.

On many of the projects I've managed, one or more engineers on the team did not know the tools or technologies we were going to use. Some might wonder how it could be that a developer could be assigned to a project who did not know the language we were going to use or the database system we would connect to or the development environment, but this happens all the time. Given the pace of change in the software industry, this is not unusual. Some friends of mine and I were joking that the only folks who still know Cobol are over 50, the folks doing Cold Fusion are in their 40s, the folks doing JQuery and PHP are in their 30s and the 20-somethings are all doing Ruby. A raft of other languages could be added to this list that would also peg the expert to an age. The point is that new approaches, languages, and tools come along frequently, and a flexible developer will embrace the new tools and continue to grow their skills. Most developers aren't flexible by choice.

As I mentioned, engineers pride themselves on their ability to make things happen in the technical world. They become adept at a particular language and wield its power like a magic wand. With just a flick of their proverbial wrist they can use their skills to make that function you needed in no time at all. They gain confidence with each use of the language. The greater their facility, the more power they have. There's nothing good that comes from losing that facility. In fact, only bad comes with that.

So then you come along and say, "Bob, you did a great job on that last project, and I want to work with you again on this really exciting project. We're going to have a pretty tight deadline, and we get to use Ruby, so you can pick up another language in the process." What Bob heard was something completely different. He heard, "Bob, blah blah blah, new project, blah blah, tight deadline, blah blah, Ruby, you won't know how to do anything and you will fail." You thought you were rewarding Bob by giving him first crack at the hottest new technology. Instead you have just ruined Bob's day. He may hide it well but he is panicking. You're telling him he has to negotiate a delicate treaty between warring parties…in French. Your choice of a new language took the rug out from under him, and he will do anything he can to get that rug back under his feet.

The first thing you will likely hear is a cavalcade of reasons why Ruby is a trendy new language but not as good as whatever it is Bob is comfortable with. I've had engineers send me lists of ISO protocol standards surrounding their preferred language and tell me that with all this standardization their favorite was the obvious winner, and I was making a terrible mistake going any other way. Mind you, my engineer had never *used* any of these standards in the past. I've had engineers print out a stack of reviews from magazines and journals decrying the new language. Almost always these are from magazines focused on the old language. The engineers will try to get me to talk to the customers to see if I can persuade them to change their minds. I've had engineers hand me a whole PowerPoint briefing to support a discussion with the client. For them to use PowerPoint, they had to be desperate.

Very little of their reactions is based on logic; it is largely emotional. Successful engineers got to this point by being able to learn quickly. They had to learn their previous language at some point and clearly became expert at it. There's no reason to believe that they can't learn this new one in short order and soon be equally proficient. All languages have the same general set of features: looping, branching, conditional testing, variable definitions, passing of parameters. The differences are primarily in the grammar. However, engineers are also very frustrated when things don't come to them quickly. They fear this. "This is stupid!" "Some idiot or maybe a committee of idiots must have written this." "It makes no sense!" are signs of this anxiety manifesting itself.

The first thing to do is calm your folks down. Start by explaining why you chose them. Odds are you chose them precisely because they had proven to be so quick to learn other technologies in the past. You chose them because they've proven to be able to think in multiple different ways before and were the best choice to lead the group in adopting this new language. They will be sacrificing some current powerful Kung Fu to have unstoppable Kung Fu in a couple of months. Your explanation will go a ways toward making it okay to leave their comfort zone on the language but what about failing on this project?

You have to reassure them that you've factored in some time for them to come up to speed, and you need to do just that. You can't assume that the same team doing a similar project but in a different language will be just as productive. You should assume they'll be about half as fast to begin with and never get much better than two thirds as fast on this one project. They'll begin by writing code that looks more like their old language, but over time it will drift into proper new language. Back in the old days we joked that when we first moved from FORTRAN to C, we wrote some really ugly C-TRAN. The same will be true for your crew. If they're moving from JavaScript to Ruby, it will start out looking like RubyScript. This will mean that even if the team is writing proper Ruby at the end of the effort, there will be a certain amount of RubyScript that will need to be refactored before delivery. As the project manager, explain how you've allowed for this. Reassure them that they won't need all that time but that it's there.

Next you have to help explain how they will learn the new language. Have you arranged for a training course, provided a web-based training base, planned on bringing in a senior Ruby guy for the first couple of weeks? Whatever you've planned, explain it to them. It's also a good idea to give them some control over their training by authorizing a budget for them to purchase any books or manuals on Ruby they think are useful.

Finally, you need to maintain your interest and support throughout. It's perfectly okay to ask if they think they're still writing RubyScript or do they think they're on to true Ruby. Let them tell you how they're feeling. Ask them which books they purchased are proving to be useful. Ask if there are any others they would like or need. Ask them how productive they feel. Are they hitting on all cylinders? Do they still feel like they're handcuffed? Reinforce that this is a learning project, and don't forget to thank them for taking this risk. It's a big one to them, and if they are making it happen, you owe them.

Communicating with Engineers

I'm a dominant extrovert on the Myers-Briggs scale. I get my energy from talking and being with people. I hate talking to people on the phone when I could just get up from my desk and walk over and chat with them. I hate email. I think it's cold and impersonal and allows people to be much harsher with each other than they would ever think to be in person. Text and chat are only slightly better. However, that's me. Your typical introverted engineer wants as few face-to-face interactions as possible. They are all tiring, stressful, and annoying. Writing software is a quiet time activity. Popping in on them completely wrecks their train of thought. That and, well, you drain them by your personal presence.

So I have had to change my ways. Email is a great way to communicate with engineers. You get near-instant response time (usually), and you get a whole lot more candor than you would if they were speaking directly to you. Email allows them to find a time when they can respond to you. A break in their coding or designing. It is not draining (though my ignorance is always a bit tiring). It is safe. The chance they'll choose the wrong word or phrase is far less when they can re-read an email before hitting send.

Engineers, even if they are feet away from each other, will chat with each other online. It is a very highly productive way for them to work. They can quickly exchange code snippets; they can share links; they can ask and answer questions. They love it.

I found that using the chat tools also worked well when I had my silly management questions.

It is hard for an extrovert to go all-electronic, but it will be appreciated by your team. I found some testers and business analysts who were E-types and got my energy that way. Everyone was happy and no one was hurt.

Engineers exchange information. Introverts find communication stressful in general, so theirs tend to be short, fact-based transactions. They use very precise language and tend not to include any extra words. This brevity can be tough, as context also goes wanting. I can't tell you how many times I've had to email back with questions like "I'm sorry, do you mean late with the task or late to the meeting?" The fact that they don't provide a lot of extra words is fortunately balanced by their dislike for reading extra words. They are very good at deducing context, so all your flowery explanations are really unnecessary and time wasting as far as they are concerned. They may even view your inclusion of all this extra blather as an indication that you truly have nothing important to do. So I suggest you keep your emails straight to the point. Skip the extra background and adjectives. Pass on your information, or ask your question and then leave them alone.

Here's a bad example of how to ask for their status: "Ladies and gentlemen, it's Thursday afternoon again, and as you recall that means I need to prepare the weekly status report for our clients. Although they've been in the daily standups, they have still requested a summary of your tasks completed, tasks in progress, plans for next week, and any risks and issues that have come up. Would you please send me a short write-up by close of business? I sincerely appreciate it. Sincerely, Dave O."

What would work much better is something along the following lines: "Folks, Need your weekly status in the usual format by COB today. Thanks, Dave O."

Business Users

First of all, what do I mean by a business user? These are the people who will be using the system you are creating or modifying. A few of the business users are the lucky ones who have been tasked as your sources of knowledge for what that system needs to do and how it should do it. They are the ones you'll be working with day to day. They are part of the project team, although they don't usually report to the project manager. These business users are the ones we are going to focus on in this section.

Business users are not all nerds. In fact, only small subsets of them ever are. However, you and your merry band of system creators have to interact with all manner of business users—I hope a lot—so a discussion on interacting with them seems important. Like engineers, there are many different types of business users, which we'll discuss a little later on. Some of the most "dangerous" interactions occur between your engineers and the business users, so it's important to prepare both sides for these interactions properly if you want to avoid some hard feelings and grumpiness down the road.

The first thing to realize about your business users is that they have "day" jobs. They have things to do that are not part of your effort. In almost every case you will find that they are not assigned full-time to your effort. Your project is one of those "other duties as assigned." Every moment they spend with you means more items stacking up at their desks that they will have to process before the day is out. They have meetings, phone calls to make, email to respond to, politics— you name it—that they have to deal with. You must be extremely respectful of their time.

The second thing to realize is that the business users who are assigned to help you are your most precious resources. They are the ones who know what *really* needs to be done to make the system a success. Their managers think *they* do, but usually they are very wrong. The business user is your best advocate for the system and, as such, your critical link to gaining system acceptance. You need to curry the favor and excitement of the business user. There is no one on the engineering side who will be able to sell the system's utility better than the business user. In fact, if the user community thinks that this new system is being pushed by the IT department, they will almost immediately reject it, even if it is the best thing since sliced bread. Users need to think that the system is theirs, built for them to their specifications. The best way to make this happen is to put your business users up front and center and hide the IT team as much as possible.

The third thing to realize about your business users is that you don't get to pick them. Keep in mind that no middle manager in his or her right mind would let their best people go off and do something that is not part of the manager's performance plan. So you are likely to get the "meh" people. This is not always the case. Sometimes those best people were the instigators of the project or sufficiently pushy to get onto the project, but in reality, your business users are generally people who could be missed. There are many reasons certain people get assigned to your project. Some of them are great and some of them are awful. You will have to work with whomever your client assigns to the project, and that's just the way it is. I discuss some ways to deal with this in later paragraphs, but either way, this is your team. Learn to love them.

Sometimes you can have some influence on the choice of business users by stating what it is you need from the user community. "I need someone who really understands all the detailed exceptions to the documented processes so that we make sure we don't miss something." That's a way to make sure you get a person who is actually hands on versus a middle manager. "The project needs someone who knows the real pain points." That might get you someone who's experienced in the current ways of doing business— again usually hands on—versus the rookie. "I also need a working-level person who is influential or whose opinion is valued." That might get you your best change agent; it might also get you the person who is most loved by management and equally despised by the worker bees, but it's worth a try. If you *don't* express the need for an influencer, you won't get one.

Another option that works best if you are part of an internal team (versus a contractor) is to go to your management and see if you can get your project and support to it added to those business managers' performance plans. It removes their disincentive to support you. Now, that the manager will still be missing an ace and will have problems in the regular operations as a result, but maybe the manager won't see that coming.

The fourth thing to realize about your business users is that they are human. They come with strengths and weaknesses just like the rest of the world. Some will be great at completing their assignments and getting things back to you on time. Some will stink at that. Some are articulate. Some are visionary. Some are clueless. Some are luddites. Some love technology. Just as we have to adapt to the personality quirks of our engineering team, we have to adapt our behaviors to the business users as well so they can make the maximum contribution to the team. One thing I will say is that the majority of your business users will not be engineers, so working with them will be a change. I hope a *"refreshing"* change.

The fifth thing to realize about the business users is that you have no control over them. Most often they are your clients, so you serve them. When they are late reviewing something, you can make note of the lateness in your status report, but that's about all you've got. You have almost no leverage, and what leverage you do have is like playing with dynamite. You can try to appeal to their sense of duty to the project. You can talk about teamwork. You can show them the schedules and how their work is on the critical path. You can try to get the other business users to apply some peer pressure or perhaps take up the slack, but if that doesn't work? Usually the only other option is to appeal to their boss, but that is not something you should do casually. In essence you are ratting the business user out to the boss, and, surprise surprise! Business users don't like that. You will have lost your best user advocate. Don't do this unless you are truly up a tree and willing to trade off completing a project that might not be accepted versus not completing the project on time. That's truly the tradeoff you'll be making.

The Cast of Business Users

So let's talk about the cast of business user actors that are sitting across from you. They are all different, and each must be handled differently. The most common roles for your production are the following:

- The manager who thinks he or she knows how things work

- The user who really does know how things work

- The user who knows what the problems are with how things work today

- The user who was assigned to this effort so he or she would be removed from the normal workflow

- The user who sees this assignment as a ticket up

- The user who joined to make sure nothing bad happened to how things work

- The user who is the management mole

- The user with the good ideas

- The proxy for the user

- The user who is the bully

By the by, some of these actors can have multiple roles. The following paragraphs discuss these business user roles and how you as the project manager should work with them.

The Manager Who Thinks He or She Knows How Things Work (The Meddling Manager)

I begin with one of the touchiest of the roles. Often a manager wants to be associated with the project. This is great. You have someone of authority on the team. This can be useful to help encourage other business users to participate regularly and complete their assignments on time. That's a good thing. You need someone who is influential to talk up the management chain about the project. You want an internal, business-side voice speaking to those more senior folks about the positive merits of the project and how well it's going. The down side is that often managers think they know how things are done, but honestly, they haven't done the job for so long that their insights are often not as useful as those of the folks under them. Sometimes the rest of the business users are cowed by their presence in each meeting. They're afraid to say, "But that's the stupid way we've been doing it for years." This is exactly what you want and need to find out. If you automate a stupid process, you get a faster stupid process but nothing more. The manager might be invested in this stupid process and therefore not realize it's stupid. Worse, the manager might feel threatened if anything changed in it. I experienced a notable exception to this once when a manager started an interview with, "You should get rid of my department and all the people under me because what we do is pointless." I was dumbstruck. She was not of retirement age; she had no new job lined up. She was brutally honest when she described how she and her staff performed a function that was largely a quality assurance task, but sadly it added almost no value. If there were flaws in what they were looking for, nothing bad happened. It would self-correct farther downstream. Unfortunately, her behavior is not usually what one sees. She will remain one of my heroes and role models.

If you are hearing more interesting information in hallway meetings with your nonmanagement business users than you are in group meetings with the manager present, then you have a problem. You can't rely on bumping into chatty users in the hallways at just the right time. You need to find a way to hear the huddled masses crying out for change. Usually you can arrange for the manager not to be present during the discussions, and you can usually be very up front about it. "Manager, I'm concerned that when you're in the room some of the folks under you are afraid to say things that might be perceived as critical of the current process or current management of the process. I don't think you'd have a problem hearing that, but as an authority figure, people can still be intimidated. Would it be possible for us to meet every now and then without you? Could you 'have something come up at the last minute' that would prevent you from attending?" Some managers won't support this, but most will.

If they won't, I hate to say this, as I'm all for complete openness and visibility on a project, but you have to go around the manager to hear what the rest of the folks have to share. Lunch in the cafeteria, a drop-in at their office, a talk while walking to get a cup of coffee are very good and very safe ways for everyone. You don't even need to let the nonmanager know what you're doing. It's natural that work would come up while you're eating or grabbing some coffee. You can start pumping the users for their ideas without them really knowing you're conducting an interview. As the outsider, you can then play devil's advocate at the next group meeting. "So let me ask a naïve question. What would happen if instead of doing X we did Y? Not knowing all the details, it seems like Y would eliminate a couple of steps but still give management insight into what was happening? Am I missing something?" Trust me, your inside source will remain nice and quiet and very attentive.

The User Who Really Does Know How Things Work (The Expert)

The user who is knowledgeable on the current process is vital to your project's success. As a rule the only people who truly know how things work are those actually doing the work. People who used to do it are sometimes brought into a project, and frankly they are more dangerous than they are useful. Things, even bureaucratic things, change. Also people's memories are not all that they're cracked up to be. So getting a current actual worker bee in the room is great. You need to pump that person for all their knowledge early on. I say "early on" because as the project progresses and changes are proposed, your most knowledgeable user can sometimes be your greatest boat anchor to change. Once they get into that resistance mode, they'll stop sharing.

A person or persons who are extremely knowledgeable about the current process and who were chosen to support this effort have a lot invested in the current way of doing business. They probably worked very hard to learn all the ins and outs of the system and may even have been given all the tough tasks because they were so good at them. They have a lot of pride in that knowledge. They enjoyed the phase when you asked them for all their smarts. Early on you focused all your attention on them and maybe even drew up some workflow diagrams based on what they told you. You wrote down everything they had to say.

But in the new world you're pulling the rug out from under them. Now you're going off "willy nilly," proposing to change everything, and suddenly their expertise is worth diddly. If you are not careful in how you interact with this person, the knowledgeable user could be a major detractor from the finished system even before you strike the first line of code. I have seen these users get frightened and start whispering things into key ears about how the new system is a nightmare—a disaster—that those techies are not listening to the users. You can have a whole grass roots movement against your system before you've even decided what you're going to propose.

To keep the current experts engaged and positive about the future system, you have to first assuage their anxiety about losing their knowledge advantage. You have to keep asking them questions. "Jill, you told me that in the current system you always have to look up the order number because the existing system can't pull up the order numbers by date. Would it help you if the system brought up a client's last ten orders in chronological order? Would that be disruptive to how the users work today, or do you think that you'd be able to teach them how to use a system that did that?" Keep asking them questions that point out how their current knowledge is needed in the future. Ask them if the proposal would handle the exceptions they know about. Use them as the expert, even in the new world.

You also need to help them feel like they are part of the change. "Joe, what I think I hear you saying is that it would improve things if the system had both the client data and the payment data on the same screen." Finally, you have to ensure that the new system isn't just as hard to use as the old one. The knowledgeable users may be afraid that not only will they lose all their invested knowledge, but they will now have to build up an equal knowledge of an entirely new set of system quirks.

The User Who Knows What the Problems Are with How Things Work Today (The Whiner)

Sadly, this person is often not the expert on the current system. In fact, he or she might be the newest person on the business user team. The issue is that your experts have no problems with the current system. They can play it like a concert grand piano. They know exactly who to call for missing data or where to look up part numbers or who needs to approve what. It's just not a problem for them. The experts were assigned to your group because they can do anything that's needed with the system. They know all the workarounds. What you also need is the newbie or, if not a true newbie, someone who is still peeved

> "That's not a pain point. That's someone's job."
> —An actual client quote

with the existing process, who hasn't been beaten down into blind acceptance by the current system's limitations.

This person might be seen as a complainer, but for the analysis phase, that's exactly what you need. Later on, when you've solved those problems, this person can turn into a strong advocate for the new system. I mean if Joe likes it and he's the biggest whiner in the group, it must be good!

Your whiner and your expert might fight with each other. They see the existing system very differently. To the expert, the system is their best friend. To the whiner, the system is the thing that keeps them from a happy life. I recommend letting the two of them go at each other just a little bit. It should flush out whether a problem actually exists or whether it's just Joe whining about some little thing.

The User Was Assigned to This Effort so He or She Would Be Removed from the Normal Workflow (The Problem Child)

I'd like to say that this is a fringe case, but it happens a lot. I understand how it happens but I still hate it. A business manager has to give up someone to support this project and still keep the lights on for the normal work. It's asking a lot for the manager to give up the ace performer. What if instead the manager could pony up William and not have to deal with him for six months, at least not full time? To the manager, this looks like a win-win. They met their management mandate to support the project, *and* they won't have to keep fixing the things William screws up! The whole team gets a vacation! It takes a heck of a manager to do the right thing and lose their best player and keep William around. So instead, you get William.

Sometimes the problem children are benign. On one of my projects, the problem child sat in every meeting like a bump on a log. He took no notes. Said nothing. Never had review comments to turn in. Nothing. Fortunately on that project I had some very good other business users, and things went fine. This is the best type of problem child to have. Everyone wins. The team gets a reprieve from this person, and the person doesn't harm your project. But there are other flavors of problem children.

Sometimes the problem child is filled with a wealth of information. Unfortunately, it is incorrect information. They will tell you all sorts of details about the current way of doing things and the tricks they know and in so doing will lead you right off the cliff. Usually you are about two thirds of the way through the analysis phase before you can peg the problem child as a problem child. For this reason I like to make sure that as I take notes on the process and proposed changes, I slip the initials of the speaker beside them. That way I can go back and validate all of the information I got from my problem child. You can't keep the problem child from talking, but once you identify their role, you know you need to verify everything they tell you.

Sometimes the problem child is just plain disruptive. They hog time in meetings or decide your agenda is stupid or dislike the document or other products you've planned. They derail or at least retard progress by arguing over what's to be done. The first thing to try is to make clear that the items on your agenda are the things that have to be delivered and that unless you are told otherwise, these are what you're going to produce. The problem child can take it up with the process people if they'd like.

You're also going to have to use all of your meeting management skills to keep things moving. You will have to cut off the problem child politely when they start a rant about something. Sometimes after repeated bog-downs with the problem child, I've had to include a rather terse statement at the top of my meeting agenda that this is the agenda we will be following and if folks have another topic for discussion, they should please send it to me before the meeting. When the problem child raises it anyway, you can reply: "<Problem Child> that's not on the agenda for today, but I'll consider adding it for the next week."

Sometimes things get so bad you are forced to talk to the problem child's manager. Remember, this is the person who dumped them on you in the first place. They know all about the problem child. I've sometimes gotten some good input on how to better manage the problem child from these discussions. You should seek out this information. In fact, that's how you should begin your discussion. "I've been having problems with problem child taking over meetings, and I was wondering, since you've been working with him for years, if you had some suggestions on how to manage him."

The User Who Sees This Assignment as a Ticket Up (The Coat-tailer)

Some people finagle themselves into your business user group because they believe that being associated with your project will be their ticket to fame and fortune and possibly a promotion. They want to ride on the coattails of this new system's success.

Sometimes that's not a problem at all. If they believe this project has that kind of impact, they will surely be a good advocate for the system. The problems arise because often these persons are office politicians and are going to play their cards very close to the chest. If things look like they might go bad, coat-tailers will turn on you in a heartbeat. "Good thing I was here, because I knew these guys would mess up!" They also don't usually want to make any particularly visible decisions about the system. They spend all their time reading the faces of the other users, especially the manager, to see which way the wind blows and then jump on that side.

Before things go bad, they may not be too troublesome. Sure they can continue to advocate for a position that the manager believes to be correct even when the real users have proven their case. They can be snippy at times if one of the business users says something that the coat-tailer knows the manager would disagree with. This is annoying but manageable.

The trouble is that as politicians they have often worked their way into relationships with various managers, and they can gain coup by whispering bad things about your project if it helps their political aspirations. They really don't care about the project; only what it can do for their status.

What can you do? Frankly, not much. Once you know that you have a coat-tailer, you will just have to try to keep up the assurances that things are going well, or, when they are not, that things will get better. Maintaining a positive, proactive attitude when in the coat-tailer's presence is about all you can do. If they still see opportunity here, they will do you no harm.

The User Who Joined to Make Sure Nothing Bad Happened to How Things Work (The Luddite)

The Luddites were skilled craftspeople of the 1800s who rejected the factory automation of the industrial age. They organized raids on factories and smashed the new machines. "Luddite" has since become a term to refer to people who reject innovation. Some people are just afraid of change. It is an emotional thing. It upsets them enormously. Some people don't want to learn anything new. Maybe they don't think they can. I've met people who just plain don't want to learn…anything. They have a job; they know how to do it; and that is the extent of their engagement with the job. They are there to pull the lever, click the button, put in their time, and go home. My interest in making things faster, more efficient, or easier does nothing for them. So they resist. They fight. They tell you a thousand reasons why it won't work. Why the current system is fine.

The first thing you have to do to calm the Luddite is to point out the parallels between their existing world and the new one. Sure, the screen will be laid out differently, but it will have the same information and, in fact, it will save them from having to go back and forth between screens. It's even less of the same thing! I've found that walking Luddites through screen shots from the current system and then showing them designs for the new system, demonstrating where everything old is now found on the new, has helped.

You also have to minimize the perception of the learning curve. This new system will behave just like the other system you were trained on. It will follow the same look and feel, and things will be laid out similarly. You already know that one.

Another thing you have to do is help the Luddites own the system. Listen to their ideas. While they might be acting like sticks in the mud sometimes, they also may save your tail by revealing the exceptions to the rules that will keep your proposal from working. Do not cut them off when they come up with the umpteenth "Well, that won't work because every six years we have a…." Hear them out, talk it through. It's an opportunity to avoid repeating a mistake from a previous effort.

The User Who Is the Management Mole (The Spy)

On most teams someone is assigned by senior management to keep an eye on the project. Sometimes it's the middle manager, but sometimes it's just someone who is a confidant or friend of the senior manager. Often this person is also playing the coat-tailer but not always. Most of the business users know who that person is. They've watched the relationships and politics of the office for ages and know that this person is "well connected." This person will be reporting outside of your chain of control to your boss or your boss's boss or perhaps even higher. Sometimes the spy is a spy for a peer senior manager, someone from an involved but not responsible organization. That's even better…not.

There's an important difference between someone who reports to a manager (we all do) and would, through the normal course of events, have reason to talk about this project and the spy. The spy has an agenda. Possibly the spy's boss has an agenda, but regardless of that agenda, the spy is not ensuring that your project is successful. That is what makes him or her the spy.

Some spies are clueless and obvious. They will very quickly talk about a chat they had with Senior Manager A. "I was just talking with Senior Manager A, and he's really excited about this project." The spy is very proud of a special status as the manager's friend and just can't keep quiet about it. That's great. Self-identified spies are the best kind.

Some spies are sharper than that. You have to discreetly try to learn the backgrounds of all the members of the business users' group to see who's who. "Have you worked with Jill before? What part of the company is she from? Oh, so she reports to Senior Manager A?" At that point, your source might continue with "…Oh yeah, they are like two peas in a pod! And you know that Senior Manager A has always had problems with Senior Manager B." If your project reports to Senior Manager B, then boom! Counterespionage mission successful. You know your spy.

Sometimes Senior Manager B is wise enough to tell you who the spy from the other department is. He or she will caution you about saying too much about x or y in front of the spy. Fun.

So now that you have a hostile force in the room, you still need to find a way to get open and honest communication. One thing to try is to make each of them state the friction points that everyone knows about. It's an acknowledgment of the dead horse on the dining table. I asked the spy to tell me what the other guys would say about problems with the spy's side of the current system. We talked about it quickly, and then I flipped it around. I asked the other folks to tell me what the spy would say about their side, and we talked through that. Then I asked the spy to tell me what other problems he saw with the system and vice versa. That tends to open up the communication between the teams. If the spy can see that the sole purpose of the project is to do better by the enterprise, then the spin on any reports going back to Senior Manager A is likely to be closer to the truth.

You also have to enforce the "no cheap shots" rule. When a problem is being discussed, you have to insist on factual discussions and no names. "The system is failing because things wait for approval for too long" is a perfectly good statement. "The system fails because Senior Manager A's team never approves things in a timely fashion" is not okay. At least while both these people are on your project, the interactions must be professional and productive.

The User with the Good Ideas (The Creative)

Often there is one person on the business user side who genuinely has great ideas. Creatives are often the persons with the least invested in the way things work today. They might be new to the company or more often new to this line of business. They are good abstract thinkers and good at grasping new concepts. They are also usually shot down by the rest of the business users. They are the ones who say, "We don't need to get two signatures; the system could track when the manager okays it and that's enough," and the old timers will respond that pen and ink are the only legal things. The creative will reply, "But there's nothing legal in this; it's just an approval to go to the next step." And the masses will say that all things are legal at all times (even though it's not so).

You have to rescue these great ideas from the maws of the idea sharks, but you have to do it in such a way as to not take an overt side. "Okay, wait, so you're saying that the approval-to-continue step is not a legally recordable event. Is that right? What's the definition of a legally recordable event?" Now the naysayers have to pony up a logical argument. If they can, fine. The idea isn't a good one. If they can't, you need to soften the blow. "Okay, so it's not legally recordable, and if we did this, we could slash two days off the processing time of getting a pen and ink signature. Are there business reasons we'd not want to have the system record an accurate approval by the manager?" This gives them one honorable way out, which is to come up with a valid business reason.

Another technique is to connect the dots of the argument to things the others have said previously. "Creative, what I think I heard you say is that picking up on problem child's idea about electronic signatures, this is a place where we could use them and save some serious time." Now problem child has to either recant previous support for electronic signatures or come up with another reason why this is unacceptable. Odds are problem child will sign on since creative's idea was "inspired" by problem child's idea.

Rescue the good ideas wherever you can. Implementing good ideas is the reason you're there in the first place. If for some reason the good idea can't be implemented now, you can still help the enterprise by documenting it for the future.

The Proxy for the User (The Proxy)

This role should never exist but so very often does. It's almost always the IT persons who are convinced they know what the user needs and will fill in for the users. They tell you that they can answer any questions that you might have about the business, or, in the worst case, they will go find out for you. They insulate you from the actual users. The problem is that they can't know what the users know. It's not their job to know it. They do not understand the context of the users' needs because they don't do their jobs. Proxies will argue that they have been the point of contact for the users for years and that they know everything the users do today. The proxy is the one who gets all the help desk calls from the users. Users talk to them about the problems they are having with the system. The proxy is the one who helps them use the system when they get stuck.

Being a proxy is nothing like being a real user. When you call tech support with a PC problem, tech support has no clue what you are doing or what else you need to know to do your job. You're asking them to help you query the system for only the customers who are 90-days or more in arrears. Does that mean that the tech support person is now an expert on cash flow analysis? No. Perhaps the greatest danger of the proxy user is the lack of context. The proxy doesn't know how a business user thinks through a problem or the most desirable way for a business user to access information. The proxy only knows what the user is doing within the limitations of the current system. That means that every conversation held with tech support is necessarily constrained by the design of the existing system, not the preferences of the business.

Proxy users should be avoided at all costs. It is not okay to let them fill in for the real business users. Sadly, many contractors are hired to build systems for an enterprise but are hired by the IT department. The business side of the enterprise is often openly hostile to the IT department. You are now caught in the IT/business civil war of the enterprise and have to find a way through this minefield. The IT department thinks they are reducing their risk by only having to talk to the end users when it comes final acceptance time, but if you let that happen you will surely fail.

In contracting situations sometimes the IT department wants to be the go-between to ensure they know what's going on with the project. Sometimes they don't trust the contractor to do the right thing or fear the contractor might promise too much. Sometimes they want to make sure that they get the credit. Sometimes the users are "too busy" to be distracted during development and can't be made available. They want to make sure that they are the face of the project. The problem is that playing telephone with subtle shades of meaning will lead to problems. You have an engineering team who likely knows little or nothing about the problem domain, an IT department who *thinks* they know, and the business users who actually do. Why would anyone think that the engineers will get the right message? This is not a good communications chain.

As the project manager you really must put your foot down on this one. Not talking directly to the actual users of the system is possibly the surest guarantee of a failed project.

The User Who Is the Bully (The Alpha)

There is almost always an alpha in the group. Sadly, this alpha is almost always the most ignorant of the problem, the least interested in changing anything, and the boss's favorite. There's not a lot you can do to take on the alpha directly. It's an approach most likely to lead to you looking for your next project without a good reference to support that quest. Instead, you have some choices. Alphas most often have these characteristics:

- Like to be right.

- Are good at taking credit for ideas

- Like to hear themselves talk

- Want to be liked (though they usually are not)

- Want to be seen as leaders (though no one wants to follow)

- Have a stronger than usual sense of insecurity

- Take failures very personally, although they do this by pinning the blame on others

So, this is a good foundation for working with the alpha. There's an old adage that comes in handy with them: "There's no limit to what you can accomplish if you don't care who gets the credit." Take it to heart. Let the Wookie win. Talk to the Wookie one on one. Exercise your active listening skills. Repeat back what the alpha is saying. Instead of the "What do you think if we did X?", which is asking the Alpha to assess *your* idea, start with "So what you're saying is that if we did X, we'd solve that problem," which is asking the alpha to assess *his* idea. Give the alpha the credit for ideas you might have had. Management knows better, and if they don't, you wouldn't get the credit anyway.

What about others on the team? Will they resent you spending all this time with alpha? Oh yeah. You need to make time for the rest of the group, too. You need to know the history of the interactions between alpha and them. They will be only too happy to tell you every stupid idea alpha had in his entire career. There are nuggets in there. They'll tell you what he's proposed ten times in the past, each time in a dumber manifestation. That's good stuff. You can use that. In fact, you can usually ask alpha about it later and get him to tell you why it was not approved. That's even better stuff.

Special Situations with Business Users

Only Managers in the Room

It seems like it should never happen, but on several occasions I have been surprised to find that I am in a room full of managers, none of whom is actually going to use the system. In one case, because of this demographic, I naively assumed that the system was supposed to support the senior managers. Everyone spoke as if the system was theirs. We were well along in the process. A set of user stories had been published, and a set of new names came up on the invite list for our next review meeting. I just assumed they were more of the same tier, but at the end of the meeting one new invitees pulled me aside and told me that nothing in this set of user stories was useful to him. He explained that the system he was expected to use would just be a burden, a second set of data to enter, and that he would be using some other tools to get his actual work done. I was floored.

This is the second worst type of proxy user scenario, and if you don't change it, you will build a lovely system that helps no one. You will fail. It's tough to change once you have begun, because if the client set this up in the first place, they believe they know what their folks are doing. More to the point, they think they know how their folks *should* be doing it better. Sadly, it takes surprisingly little time for managers to forget the perspective of the person in the trenches once they start spending all their time at the next tier. Sometimes management doesn't want to bother their people with thinking about how a new system would work. They want them to continue doing the day-to-day work and not get distracted with this new potential world.

You have to be blunt and say that you need real-world users in the room. Sometimes you can appeal to the dictates of process. Process documents almost always have something about talking to the actual end users. If they don't have any process documents, pull up the Project Management Body of Knowledge (PMBOK) or something from the Software Engineering Institute (SEI) or some other authority the client respects and help them through their hesitancy to include the actual users.

You might try to persuade them by talking about how inclusion now will help with acceptance later, that this is part of the change management process. Managers like to think they are being proactive on change management. Change management has cachet. You can also try the weedy details argument, which is that as a senior manager there may be some in-the-weeds types of things that they don't have to know about anymore but that could affect the utility of the system. Managers like to think they are big-picture types and have other people to remember the pesky exceptions and details.

In a worst-case scenario, you might have to skulk around the halls and talk to the actual users on your own. This is a very risky approach but sometimes is your only option. It's risky because it's hard to explain how you came up with your new ideas on the system. How did you suddenly know that electronics products have to include either the UL or the European CE information or that volume sales discounts have to be approved by two tiers of managers? You can use the "divine intervention" excuse only so many times before even the densest manager smells a rat. So only do this independent exploring if you've been turned down cold on talking to the users.

No One Knowledgeable in the Room

Other times I've had a group of people who are the target demographic and will be the users of the new system, but I'm having a tough time getting answers to what seem to be the simplest of questions on how things work today or what problems might exist. Here's a variation on the problem child scenario. You have nothing but problem children and newbies. No one knows anything useful. The newbies—well, they are useful for new ideas once they get a handle on how things are done, or at least what should get done. I've talked about handling the problem children, but what do you do now?

The first thing to try is to go very delicately to your project sponsor to see if you can get some other folks in the room. The tricky part is that your project sponsor picked these folks, or at least accepted them. You're asking project sponsors to go back to their peers or their superiors and ask for additional or different people. They usually balk at this. Also, your sponsor might have no idea what the actual requirements are. To the project sponsor, these users seem very knowledgeable. You seem very picky and difficult.

Another option is to pay very close attention to the conversations between the business users. They usually keep mentioning someone with whom they need to check or someone who did that in the past. They will likely mention the person you really need to talk to. Sometimes you can get by with using your assigned business users as go-betweens to the actually knowledgeable person, but that's not an ideal solution. Asking if this person should be invited to the next session often works. It takes the action item to talk to that knowledgeable person off their desk, so it's actually easier if he or she is in the room.

Sometimes the reply is that the person is too busy to be part of your effort. If that's the case, you just might discover yourself in their office chatting about the project and thanking them for their indirect support.

In other cases, when no one name was consistently mentioned, I have just wandered into the area where folks do that task and started chatting up people in the halls until I found a likely victim. Once I even got caught doing this by my unknowledgeable folks, but it's easy to say that I just wanted to see the actual work being done to have it gel in my mind. They chuckled at my simple wits and moved along. Phew!

A Word of Caution About Fights Between Business Users

It is not uncommon to see some intense emotions when a group of business users are discussing how they might change their world. Like all debates, some of this is great and useful for fleshing out issues and potential problems. Some if it is just plan scary. I have been on projects in which every meeting included a screaming match between two of the business users. Unlike arguments within the engineering team as the project manager, the business users don't usually report to you. You support them. So my advice is not to get involved. You can't possibly know the history there. They might have hated each other for decades before this project began. One might have screwed over the other on a previous project. Maybe one got a promotion over the other, or their department got the cool new something or other. Whatever it is, odds are when they are arguing about functions for your system, they are *not* arguing about functions for your system. They are arguing about old baggage that has nothing to do with your project. Do not pick sides. You can't win with either one and you could end up with both of them turning on *you*.

If the arguments are getting in the way of the meetings, then you may have to go around to your project sponsor and see if something can be done. I can't say this has worked well for me. I usually get the "Oh yeah, those two have hated each other ever since the re-org." You may just have to endure their hatred for each other and plan agendas with an unscheduled but predictable screaming bout.

Business Users and Engineers: When Worlds Collide

The engineers really need to hear and understand what the business users want from them. They need to hear what the problems are, and they need to understand the context of the problem. They need this so they can then use their creativity to come up with innovations that make the system a worthwhile endeavor. Unless you're just replacing a system that is obsolescing, the goal is to achieve some sort of improvement by creating or integrating a new system. The users usually don't know what the technology is capable of and therefore won't be able to ask for a feature directly. When they do have ideas for what the system could add, they often talk about an incremental improvement. If you want something truly revolutionary, you need the engineers to chime in. You need the engineers to conceive of how the technology could work in the users' hands to make their jobs better, faster, or more reliable.

Reading even the best written user stories or requirements statements does not necessarily make this happen. Engineers and users have to get into a room together. Although your engineers are mostly introverts, the end users are usually a wide mix of personality types. Your engineers are logical system thinkers. They want to optimize something. Your users are mostly process thinkers. They execute the process they are responsible for day in and day out. There can be problems if you don't prepare each group for talking to the other. You have to be the glue that connects them.

Prepping Your Engineers for the Business Users

Before your first meeting with the business users, it's a good idea to pull your engineers together and talk to them about what is about to happen. This is a big deal for them. They are getting to meet the people who will be using the system they will be building. This is the audience the engineer wants to delight and impress. At the same time, this is a group of strangers who do not behave like engineers and are

quite possibly extroverts who will be draining and exhausting to talk
to. If you have any folks who are even near the Asperger's scale you
have to be especially careful to prepare them for the encounter. They
are genuinely unaware of social signals and can easily run afoul of one
of your business users. What follows are categories of items you
should consider talking through with your engineers.

Explain About the Audience and Their Role

The first thing is to establish that the business users are the owners of
this system. These are the folks whom the project needs to delight.
They are why we, the engineering team, are here. There's nothing in
this for us; we serve them. If we get unhappy users, we will also have
unhappy engineers. The users are the subject matter experts; *experts*,
on what needs doing. You should stress the word. Remember they
respect expertise. You have to tell your engineers never to assume
they know better. It doesn't matter what is written down in some
process document. If the user is saying something different, odds are
the user is right. No one keeps documentation up to date. The
engineer's job in every meeting is to learn as much as possible about
what the users need, what their jobs entail, and how they do them.
The users' job is to try to explain all that to you and the engineers.

Explain About Skills and Smarts

Engineers generally find conversations with nonengineers stressful
and frustrating. Talking with anyone drains the introvert, and now
they have to talk to folks who at the very least are not as technically
gifted as they are and in some cases aren't as bright. They aren't stupid
people, but they might not have a 130+ IQ either. Recall that
engineers most highly respect intelligence and mental agility. If you
have users who are good at their jobs but do not possess a superior
intellect, you run a very high risk that the engineers will snub them. In
the engineer's mind this person is of lower caste and can be dismissed.
You have to explain to your team that the users might not be as quick
or as smart as they are but that the one thing they *do* know more about
than the engineering team is how and why they do what they do. It
makes no difference how large their brains are to their qualifications
as subject matter experts in their own business. The users, by contrast,
probably will be shocked at how little the engineering team knows
about the actual business of the enterprise.

Explain About Language

You should talk to your team about how the users will likely use language differently from what they are used to. Engineers can get annoyed with users because they are not terse, direct, and precise. The users talk all around an issue without ever mentioning it by name. They are sloppy with language. They'll use a word with a known meaning in ways specific to their work but contrary to the common meaning of the word. This is very unsettling to the engineers. It's a violation of the rule that words have understood meanings. If you do not prepare your folks in advance, they will try to correct the user. You will find them pulling up their smart phones and showing the users the dictionary definition. They will offer other, more suitable, words. This will irritate the business users. Ask your guys to resist the temptation to correct the users. In fact, you also have to ask them to parrot back the words they are hearing, even if they are being used incorrectly. Use their definitions, as painful as it might be to do so. They'll feel more comfortable that you understand them if you use their words. They'll therefore like the system you're building for them and love you for doing it.

The business users are likely to talk in flowery, extravagant ways. They waste words. Engineers bridle at this. Where's the meat? Where are the facts? You have to tell them to relax and let the folks we need to please have their moment. Remember, the users are pretty excited about this whole thing, too. They've been waiting for some improvements in their systems for ages. It probably took years to get the budget approved, so give them their moment. Let them wax eloquent. Pay attention though, because there might be a nugget in there somewhere. Also, they've never met you guys and they know that you're the ones who are going to be bringing this to life. They're excited to meet you.

Never Say Anything Judgmental

An engineer can spot something foolish a mile away and in about a half an instant. That's great; it's how they remove problems from systems, but you have to caution them about calling a stupid thing "stupid." You have to tell them never to say something's stupid or broken or lame. You're in essence saying that what the users do is stupid, broken, or lame. Instead say, "Do you find that difficult or tedious, or would you prefer to be focusing on something else?" "If the system could let you look that up, would it make a real improvement or not really address the main problem?" You can say things like "That sounds hard" or "Would you want an easier way to do that?" but not "Wow, whose idea was *that*?"

Explain About Technology Awareness

Engineers know their technology, and good ones know great ways to apply tricks to solve real-world problems. The business users do not. I have seen the engineers roll their eyes at the technical naivety of business users. They can then start to question the intelligence of the users and even start to bully them. This can't happen. You have to explain that the engineering team is here because our job is to provide the engineering talent. To solve *their* problems. The users are experts at their business but won't know the first thing about technology. It's not part of their jobs. They process orders; you do technology. Try flipping it around on the engineers: "They think you're pretty lame at doing orders, so don't think they're lame for not knowing what sorts of searches are hard or easy."

The users won't be able to see revolutionary changes. They are in the weeds doing the job and are wishing for incremental improvements. Don't let your guys disparage those. Make sure they know to let users voice their ideas; often a couple of these will lead the engineering team to that revolutionary change we're looking for. Hear the users out!

Explain About Patience

Make sure your guys know not to jump in too quickly with their ideas. You want to hear the whole process before you do or you might be optimizing only a subprocess instead of the whole thing. You could be derailing the user's explanation so we then miss things. Tell your team to make notes to themselves about their ideas and pick a time when they are surer they know the whole business landscape before presenting them. The users might feel you're cutting them off, or they might get too scared if you are proposing changes before they've had a chance to explain about their world. Now you're pushing things on them they never asked for. Let the system be theirs. Ask your most likely idea sharer how he'd like it if I turned to him and said, "Bob I understand that you need to get to work in the morning. You should buy a blue Honda Fit at Springfield Honda." I have no idea where you live or if you like to drive or if a bus or walking would be more convenient or if parking is easy here or at home or if blue is your favorite color. I might be right, but why would you accept this since I've not proved I understand your transportation needs.

As an aside, I was going to use the example of "Bob, I understand that you are having problems with your .Net application; you should use Linux." However, if you say that to an engineer, he or she will completely miss the point you're making and instead go on a very detailed and long-winded explanation of the superior features of the preferred environment. So I went with the car example.

Tell your folks to wait until they've let the users get winded before they suggest changes.

Explain About the Users' Feelings of Importance

Often the users are very proud to be part of the effort. They were chosen for this job because they know the most about the problem (or so we hope). They were chosen because they are leaders. They were chosen for a number of reasons. They might have a strong sense of self-importance. Your guys might not see it this way.

You have to explain how the users might feel to your engineers and make sure they give them time to calm down and return to ground.

Another angle on this is that most users think that what they do is unique. No one does what they do. You don't want your guys to burst this bubble, because there's no winning there. Tell them not to correct the user if the user goes off on how they are the only ones who do this even when the engineer can name three commercial products that do the same thing. I did have a guy say, "Really? Then why does Acme have a product for it?" I groaned. You have to explain that being right isn't always the right thing to be.

Explain About Socializing

Socializing is harder for the engineers to understand. There's a desire by the extroverts to connect with people. Debra Tannen's book *"You Just Don't Understand"* asserts that most women use conversation to connect more than to pass facts. Engineers are mostly men, but when it comes to conversation, they are even more fact based than average. Warn your folks that the users are excited to meet them and may want to know all about them. The engineers might not feel a need to learn about the likes and dislikes of the users, but some of the users will be more comfortable once they know more about the engineers than just the details of their occupation. For them it's how they learn who you are and if you can be trusted. If you cut them off cold, they'll retreat or not share the key information that we need. Tell your guys to be brave and try to play along. They don't have to bare their souls but try to make some small talk if they can.

Explain About the End of the Meeting

Engineers, if left to their own devices, will rise from their chairs at the end of a meeting, pick up their notepads, and walk out the door. The business is done, facts have been exchanged, and we're outta here. You have to explain that these meetings are significant for the users and that there is an expected protocol for ending them. Make sure they know to thank the users for their time. The engineers will look at you oddly. "Of course they have to spend time in the meeting. How can they expect us to build them a system if they won't commit some time? This is madness." They are right in that sense. But you have explain that the engineering team serves the business users, and they've just spent an hour of their day on this side project while their day job work piled up on their desks. Also, everyone likes to be thanked. It's just a nice thing to do.

Prepping Your Business Users for the Engineers

If you have a group of business users who have never worked with a merry band of engineers, then you truly have to prepare them for the fun. Sometimes this is difficult because the first time you meet them the engineers are already in the room, but you should make every effort to sit down and talk to them about the engineering team and their special quirks.

Explain About Their Skills

The first thing to do is explain the credentials of the team. You have to make a case for why the rest of the baggage is worth their tolerance. I'm not talking about a hard sell here, but a simple "The team is so excited to be working on this project with you. I've gotten everyone I wanted to be on this team. Bob is an absolute whiz with a database, which is something that's going to be a major part of this project. He might be a little odd in meetings, but boy, can he deliver. Betty is painfully quiet but is an ace when it comes to the user interface…"

Explain About Introverts

You also need to tell your business users that you have a team of introverts. That they are brilliant but quiet, shy, don't talk a lot, and are very logical almost like Spock on "Star Trek." For all their excitement about the project, they are also really nervous around people, so if they look awkward, well, they are.

Explain About Language

You should explain that your team is very detail oriented and fastidious, which is perfectly suited to writing software. That said, they can't turn it on and off. They are very detail oriented and fastidious in their use of language as well. It might even seem like you're talking to a computer at times. They tend to spit out facts or ask questions in their odd manner in as few words as possible and then go quiet. They're listening; they just don't tend to add many adjectives or use one more word than is necessary to convey their meaning. You can also share with them that if they start to hear "Star Wars" or "Star Trek" references, they must trust you. They're relaxing.

Along with this fastidiousness is their precision in language. If the office uses a word to mean something different in this context than in its general use, I promise you, you will see an eyebrow go up or a weird look. Remember they are always trying to be mercilessly precise when instructing the computer to do things, so hearing uncommon meanings for the same word causes them distress. If a word is used for something specific in your office, just let them know what it means in this context and they'll calm right down.

Explain About Greetings and Handshakes

Many an engineer has mastered the dead fish handshake. It's perfectly okay to warn your users that your guys are brilliant but a little weak on the social graces. Ask them if they could please be understanding about getting little eye contact, a clammy handshake, and some grunts for hellos. Most engineers are shy beyond shy, and it's just who they are. I recommend sharing the joke "Do you know how to tell the extroverted engineer in the elevator? He's the one looking at *your* shoes." That pretty well explains things.

A Quick Rant About Process

Before we go further into the project, I want to talk a bit about development processes. What I'm about to say might upset some people. Some organizations might take such offense as to revoke my certification with them but here goes…Process is not the solution. Period. In fact, it often gets in the way. There! I said it.

If having a fully mature process could remove the risk and high failure rate of software projects, then we would have solved the problem with Structure Analysis/Structured Design or Information Engineering in the 1980s. The Object Modeling Group would have delivered us from risk with Object modeling and design and Unified Modeling Language (UML). Any number of vendors would have solved it with their 4GL tools like Access or Dbase III. Heck, we would have solved it with the classic Waterfall. The Software Engineering Institute (SEI) would have solved it once a company gets to perhaps level 3 on the maturity scale. The Project Management Institute (PMI) would have solved it by certifying the project managers. But we've had all of these, some for a long while, and software is still at high risk for failure. Process is not the solution.

I am not impugning the intentions or accomplishments of any process organization. They are trying to document things that have worked well for a number of projects. Process has a chance of reducing the incidence of common

> Process leads to project success at about the same rate as cookbooks lead to gourmet meals.

mistakes…sometimes. Proponents of processes sincerely believe that if you follow this process you will have success. Process leads to project success at about the same rate as cookbooks lead to gourmet meals. And why is this? Because knowing *what* to do is very different from knowing *how* to do it and—even more important—when to do it differently. Process cannot make up for a lack of craftsmanship.

Enterprises don't have an army of great chefs. They might have one or two, but they need to make meals for hundreds. So the siren song of the process people is appealing. Process tells managers that failures on projects are due to the lack of following a mature, complete process. Managers will establish the process group; sometimes it's called the process improvement group or engineering improvement group. Either way what they are trying to do is write and then enforce a more perfect cookbook...and it's not helping.

What's worse is that *every* process group I have ever worked with has eventually morphed into an inward looking, self-serving organization. Rates of adoption and compliance with the prescribed process became much more important than actual results. Any failure on an individual project was ascribed to a deviation or poor completion of a process step. I argue that you can be a Capability Maturity Model Integration (CMMI) level 5 company and still suck eggs. You would have very good, repeatable processes for sucking eggs, but you still suck eggs. The process department eventually manages what becomes a paperwork drill. Did you create the artifacts that the process dictates in a timely fashion? How many hours did it take? Were the artifacts complete per the standard? The process has become the self-licking ice cream cone.

If you believe that I am therefore advocating anarchy and randomness, you are incorrect. There are a set of steps you need to complete for every project. There are some better and worse ways to do them, but as with cooking, following the process to the letter is not the goal. A delicious meal is the goal, or, in our case, a system that meets the needs of the business. Perfection of any given step is not going to improve the odds of success. When gathering requirements, it's important to try to include as many of them as possible, but the reason for having a complete set is to avoid the very small probability of missing the one requirement that causes a massive change to the architecture.

What are the odds that such a requirement exists that could remain hidden yet cause such a massive change? We are perfectly comfortable hitting a section of a textbook that says "the proof is left to the interested graduate student," but our process dictates that we nail every minute detail down into our design documents. Ignoring the fact that the requirements will change, the environment will change, and the team will change, our processes dictate robust completion to move on to the next step.

What I advocate is sufficiency. Each step must be completed to the appropriate level of completeness and thoroughness to support the next step. When are you at that level? That is where experience and craftsmanship come in. It is from experience and skill that you can determine if you're done. The process tries to remove the need for skill by dictating a lock-step approach that even a child could follow, and this is why it fails. A child cannot lead a software development team any better than a child can write the next great novel.

What follows here are my suggestions and approaches to the steps that are most often required on a project. I urge each of you to adapt them to the unique circumstances of your project. If you can't get your users to figure out how to write down their needs, don't. Change it up. Use a screen mockup, a whiteboard, a spreadsheet, whatever, to help them through it. Go to rapid prototyping. Do what makes sense to be able to deliver something useful for your clients. Don't keep pounding your head against the wall of the mandated deliverable.

Process Departments

Many organizations include a group of people who ensure compliance with the approved way of doing business. They probably were the same ones who originally wrote down the approved way of doing business. In better organizations the authors were the most successful and more experienced practitioners; however, in a surprising number of organizations the process team is staffed with people who stopped actually practicing their craft long ago. Now they do meta-engineering by describing the perfect process for doing work the way they believe they generally remember. I have a separate rant about process later, but for now, suffice it to say that I am a fan of as little process as possible. The process folks do not share this view.

The process department is there to monitor your compliance with the standard, whatever that might be. They will usually require you to hold some separate reviews with them so they can measure your project's adherence to same. If you think testers are process people, you've not met the process people. They wrote the rule book and, by Jove, they are going to make sure people follow the rules. Depending on the organization, the process department either rules with an iron fist or is moderately ignorable. In very few organizations is it somewhere in between. Your project is another test of their dominance. If they can bend you to their Tao, then they are happy. If not, they continue to feel weak and impotent. It's easier for you if they are happy even in organizations where they are weak.

At the beginning of the project, you will almost always get a visit from the process people. They will want to make sure you're following the startup processes and know where the Repositories of Goodness are. They are generally eager to help. Let them. Let them teach you about the processes, the checkpoints and gates you will have to navigate, and the people who have to approve things. Do not buck the system any more than you absolutely have to. If they rule with an iron fist, you will lose, and if they are of little influence with senior management, they might start whispering in various ears that your team is a bunch of cowboys who won't follow best practices and then just wait until a defect appears and jump all over you. That is how they tend to get an iron fist. They keep pointing out flaws in a system that adherence to their processes would have prevented. I believe they are incorrect, but their pitch sells well to senior management, so work with them. If they want to have a meeting to review something, hold that meeting. If they want a write-up on how you came up with the decision to build versus buy, go ahead and write it. The time you spend trying to get a waiver from an approved process isn't worth the hour it would take to document your decision process. It won't take much time, and it permits them to show senior management how much value they are adding.

As a rule, the process team is looking for your willingness to meet the spirit of the law, not the exact letter of the law. On an MIS project I was working, the process dictated a detailed description of how each component of our system would map to the classic OSI seven-layer networking model. Such a description truly would help no one. The process people had read an article about how this description was a great idea, and suddenly it was a requirement. No one actually needed it. In our case, it would reveal no insights.

After I had time to vent quietly somewhere removed from anyone's ears, I popped over unannounced to chat with the process team member who'd been assigned responsibility for our project. I explained that I was looking over the next section of the process for our project and came upon this requirement. I told him how our entire application was going to live at the application layer and the component map would be unenlightening. I asked him if there was some other information he needed that *would* be of value. He thought for a moment and then asked if I could map out the servers I'd be accessing instead. In two minutes I (1) avoided developing a piece of useless documentation, (2) escaped fighting with the process department for weeks on end, and (3) got to reuse a chart I would have to produce for the security review, help desk, and operations teams anyway. Poof! Full compliance and a happy process department.

Risk Management

There is a lot of good that can come from risk management but, in my opinion, not nearly as much as there is fuss about risk management. The thinking exercise of taking time to review the project and its dependencies, the organization's operating model, the staff, the infrastructure, the vendors, the degrees of certainty on things, and how they might affect the success of the project is wonderful actually essential. I learned a saying from my dear friend and mentor Stan Lucas, which is that "In project management, there is no such thing as a good surprise." Risk management takes a crack at reducing the number of surprises.

<Heresy Alert!>

I believe that separating risk management into its own subarea of project management with its own certifications, processes, positions, and the like is fundamentally flawed. It separates tracking and managing risks from the management of the project. Risk management is just another set of tasks to track in your project plan. The project manager should be tracking status and managing the resources assigned to risks just as he or she would every other task on the project plan. The textbook processes currently out there prescribe a risk management approach that is a sidebar set of meetings held once a month (at best) by an overlapping but not necessarily identical list of people from those accomplishing the project. This makes no sense.

Some tasks in a typical project schedule are risk reduction tasks, but we think of them as normal processes. We test a system to reduce the risk that the developers made an error of such a magnitude that the system can't be used. We train users to reduce the risk that they won't be able to figure out the system on their own. We write manuals to reduce the risk that users won't remember everything in the training. So why do we separate other risks into a separate set of processes and people?

You would not have one person drive a car and another person ride along as a safety officer. The driver is responsible for driving the car in a safe manner. Some large, high-risk production companies have actually eliminated the position of safety officer. Why? Because they found it created a culture in which one group was responsible for producing product while a separate group was responsible for safety. Accidents and injuries actually went down when safety became the duty of every individual on the production line.

So risk management is just another set of tasks on the project schedule, another product that is frequently updated and reviewed and should be performed by the project team members. Projects need to look at anyone with an assignment on a risk task as part of the team.

With that said, how does one manage risks? How does it work? Well, the mechanics of it are actually pretty simple. The first thing you do is write down all the things that might happen to your project. Things like the following:

- staff not coming on board on time to meet the schedule

- an external project on which you depend being late

- equipment not being up to the load you're going to place on it

- clients not completing their reviews of key documents on time

- the team not coming up to speed on the technical environment within the window provided

- customers being unable to agree on the requirements for the system in a timely fashion

- clients being unwilling to agree on the priority of requirements

- holiday and summer vacation schedules making team or reviewers unavailable during key phases

- recognizing policies that must be changed before the system can move ahead

- users being unable to use the system with sufficient efficiency to gain the productivities on which the ROI was based

- other projects taking away needed resources

- being faced with upcoming strikes or labor negotiations

- changes to the infrastructure that could affect the project

- vendor support not being reliable

- you name it

If it could affect the success of your project, it should be included in the initial list. I believe that unless your project is in a war zone or a flood zone, you can omit listing the destruction of the data center, but otherwise you should list every possible risk that could come up.

There are a number of reasons to do this.

1. When you review them with management and the customer, it sometimes spurs action to head them off.

2. It lets everyone know what the consequences of deciding *not* to initiate action to head them off will be.

3. It provides a meeting at which you can remind management of the risks and consequences of not initiating action to head them off.

4. It provides a set of documentation in which you can prove that you told them so.

So now you have a list of things that could go wrong. You could go ahead and come up with ways to prevent them all or figure out what you'd do if they happen or try to reduce their impact if they do happen. Odds are you don't have time, and possibly more important your management group probably doesn't want to wade through them all. So there's a standard drill in the texts in which you estimate the probability of each risk and the impact on the project if it happened. You then focus your attention on the most probable, highest impact risks. That makes some sense to me, but later I'll explore some reasons why you would want to elaborate on some other risks that fell into the lower probability/impact categories.

Probability/Impact	Low	Medium	High
High	Yellow	Red	Red
Medium	**Green**	Yellow	Red
Low	**Green**	**Green**	Yellow

Table 1. Classic Risk Probability and Impact Matrix

You need to use your best judgment regarding which risks are most likely. There's no mystery here; it's simply your best guess. Most risk management processes have four brackets of probability. Some have three, which I like because high, medium, and low is painfully simple. Many write-ups try to apply more rigor by assigning probabilities to the categories: for example, high means 85% to 100% chance of occurring, medium is 50% to 85%, and low is everything under 50%. This is a fool's errand. The exact numbers are irrelevant. You rarely have an exact figure for the probabilities. What you have is your experience, your insight, your gut, and frankly that's good enough for this exercise. I have attended meetings during which a group of people debated whether the probability was really 85% or closer to 80%. Honestly, what a waste of time! When we were little, my older sister and I had imaginary bears that lived in the imaginary caves in the rocky slope right outside our kitchen window. We constantly fought over whose bears were better (for the record, mine were better). Arguing about whether a risk is 85% likely to happen or 80% likely to happen is equally unproductive. You'll hardly ever have any science or hard numbers to back up your probability.

Impact is also typically just one's best guess. Some of the answers are easy. If the staff are not on board, then you cannot start, and the project has a day-for-day slip in schedule. Some are harder, for instance, what if you get half the staff? Would you actually make half the progress? Odds are you'd make less, but how much less? If you're missing the person responsible for getting everyone access to the sensitive documents, you might not get anything done.

Categorizing the risk to schedule, cost, and quality is fine. It helps management feel better when making some decisions, but it isn't all that useful to the project itself. Furthermore, risks rarely fall into only one category. If the schedule slips because something didn't get delivered on time, then your cost goes up too since you didn't fire the development team. They sat on their thumbs for a week. You had one week of unproductive effort.

Your engineering team is great at finding risks. They will come up with everything under the sun. Remember, they are much more literal than you, much closer to the building of things than you, and have been delayed before by things outside of their control, which they hate. They remember these problems and will lob them back to you. Some will be way over the top: Meteors might take out the building! Wild goats might storm the data center and erase all the backups! But seriously, they are always trying to control their world, and now you're talking about the world they love the most. They will tell you about all the upcoming releases of underlying products that might not be compatible. They've been reading about new protocols, deprecated functions, library problems, hardware incompatibilities, firewall limitations, and every other thing under the sun. Seek this out!

The engineers are also an amazing source of solutions. So often we say things that our engineers take as direction, and it's only because we don't know what we don't know. For example, we might talk about using the standard software suite in the building at the time. The engineers know of a problem in trying to do what we need to do with it and will tell you about the risk. You panic and say, "Oh my goodness! Whatever will we do?" They will reply that there's nothing *to* do given the current suite of software. You look despondent. Then they might add that the upgrade to the new version doesn't have that problem. *What*?!! You mean it's that simple to solve? Why didn't they mention it? Well, you said we had to use the current software suite. You told them it had to be the current environment and they inwardly groaned. They might have outwardly rolled their eyes, and you completely missed it because they always roll their eyes at you. Given that they are usually looking down at the table or at their hands when speaking, you just never get to see it. They marvel that you can get yourself dressed in the morning given your lack of knowledge and technical acumen. They never said anything earlier because, well, you didn't ask and you're less annoying than some managers. Actually, the odds are good that they did mention it to you but you just didn't understand what they were saying and nodded your head knowingly and moved on.

This interaction is common, and it has a lot of similarity to talking to your aged grandmother. Her hearing is going or maybe mostly gone. She misses about half of what you say but is too embarrassed to ask you to repeat everything you say, so she just smiles, laughs if that appears appropriate, and says things like, "Oh, that's nice." You know Granny hasn't understood you, but she's so nice you let it go. The engineers know too that you didn't get it. You're too embarrassed that you didn't to ask them to explain it. The engineers know you won't really get it, so they let it go.

One thing that I've found very useful that I don't see in every risk management process is to document how you know if the risk has happened. That sounds obvious but actually isn't always the case. Obvious ones are "Company has a general strike preventing you from accessing the facility and appropriate personnel." Yes, easy, we know it because there are picket lines, building is closed, CNN is out front, things are bad.

There are other risks for which you can set clear criteria to determine if they are happening. For example, "Equipment is not delivered in time." Think about it and set a date by which the equipment should be here, and then you know definitively whether it's happened.

The occurrence of many risks is harder to confirm, however. For example, a common risk is "Stakeholders do not make time available to meet with team when needed." If one stakeholder is not available for one meeting, has the risk happened? Odds are no, depending on which stakeholder couldn't attend. If none of the stakeholders is available for any meetings, has the risk happened? In this case, the answer is an obvious "yes." But what about anything in between? When would you go to management and declare that you are not getting the support you need to make the project a success? It's important to think this through ahead of time. You'll be up to your eyeballs in problems and a healthy dose of problem denial once the project is under way. Are there decision points in the project by which you'd have to have the stakeholder input? Are there document approvals that would have to have their signatures? Are there design phases that need their blessing? List these now, early in the project, while you are still able to think clearly.

Another example is "Staff is not assigned in a timely fashion." If no one is on board, then sure, the risk has happened. What if you have one position unfilled on day one? Do you raise the red flag? What if you have the whole team, but they are still mopping up items on their old project, so you don't have them 100%? What if you planned on persons A and B for 50% each but instead got 100% of person A? What if it was person C?

What about "Learning curve on new tool might delay timely or quality completion of code." How would you know if this has happened? How can you determine whether they're behind because they're behind or because of the learning curve? What are the signs that they aren't comfortable with the tool? Is it that you see them poring through the manuals all the time? They're on the phone with technical support? They're cursing the name of the manager who made them use this stupid thing? It could be as simple as 100% of them have not gotten their certification in MontyTool. Whatever it is, write it down at the beginning.

We've listed the risks, estimated how likely they are to occur, estimated or perhaps calculated their impact on the project...big deal. The most important thing about risks is to answer the question, "Can I do anything about it?" Here's a simple truism: If you can't do anything about it, you have just accepted it. You can monitor it, recognize it when it happens, but there's nothing more to do in terms of risk planning. Move on to those that you can do something about.

Risk management has become its own self-licking ice cream cone, so what you do about risks has all been categorized. I've seen multiple lists of them, but practically they fall into five categories:

1. Prevent it from happening—the first thing you should always try to do. Make it go away.

2. Avoid it; do something that makes the risk a nonfactor for you. For example, if you have a risk that you might not get enough of

resource X and can find a way to redesign your project so it doesn't need resource X, then you've just avoided the risk completely. Ole!

3. Reduce its impact if it happens, also known as mitigation—second option if you don't have a way to eliminate it or avoid it. Do something to control how much of your world it affects. Maybe you can't eliminate the need for resource X, but you can isolate it to the reporting function of your system. Everything else will be fine, but your reports will be delayed or perform slowly.

4. Transfer the risk to someone else. It's kind of a low blow, but if you can't prevent it, one option is to stick someone else with it: "an action assigned is an action completed." For example, if you have a risk that your operations staff's lack of experience in fault-tolerant systems could lead to outages and client dissatisfaction, you could contract with a cloud vendor to host your system, so that now the risk is the vendor's. The risk still exists, of course, but it's no longer yours. The response and consequences are built into the service level agreement you sign with your cloud vendor.

5. Accept it; suck it up. If it happens, it happens, and we'll deal with it. Meteor strikes on the data center fall into this category. There's nothing you can do to prevent it. You could build a completely redundant data center and sync all your data in real time, but the cost would be astronomical. You could contract out to a cloud vendor, but if fear of meteor strikes on your building is your only reason for doing so, few reasonable CIOs would sign the purchase order.

I believe it is an academic effort and a waste of time to categorize the risks. Regardless of the category of risk, you are going to decide what actions to take, assign people to those actions, track the actions through completion in the project plan, and report on them. There's no difference. Time has been lost in meetings arguing about whether the risk affects quality more than schedule or cost. Who cares? It's bad and you need to do something about it, so get focused on that.

Some people get confused with risk response plans. You always want to spend some time thinking "What would I do if the risk happens?" If the equipment does not come in on time, then I'll use the old stuff until it does, or I'll rent equipment until it does. If the stakeholders don't play ball, then I'll call a halt to the project, engage with the sponsors, and hold a summit to discuss it until they commit to the schedule. If the team is not learning the new tool fast enough, then I'll contract with some outside experts to fill in until they do. If I can't get approval of documents in time, then I'll either delay the project or seek sponsor approval to move ahead with the draft documents. Decide now what you want to do before the project vision cloud obscures your ability to decide. You'll want to please those clients and ignore the problem. "Damn the torpedoes! Full speed ahead!" But you will fail. Instead, decide now, when the project is clear and sane, what course of action you will take and trust your decisions later. Listen to the calm you.

Problems Versus Risks

There's a lot of attention paid to the difference between problems and risks. The textbook definition of the two is that *risks* are problems that might occur, and a *problem* is something you're facing right now. So, the difference is only temporal. For the process people, this is an important difference; for the project manager, not so much. A problem might be the nearest tiger you need to shoot, but all of those other tigers are coming to eat you as well. The probability of a problem is 100%, but otherwise you need to do all the same steps. It's another reason why I prefer to include risk management into regular project reviews and not hold separate meetings solely to look at faraway tigers.

Risk Management Process

The texts talk about how you need a risk manager and that you need to hold separate risk management meetings. Again, the self-licking ice cream cone of process, in my opinion, is wrong on this. The problem with holding separate meetings is that it separates risk management from the project. People will blow off a risk management meeting but not a monthly project review. I recommend incorporating a review of the risks into each project review. In fact, I recommend including an update on any risks in each weekly status report. When there are tasks to which people have been assigned to eliminate, avoid, reduce the impact of, or reassign a risk, then these efforts should be included in the project plan. The tasks need to be reviewed, their status checked, and any problems raised right away, not at a monthly, ill-attended risk meeting. Risks can't be separated from the rest of the project. Often a senior manager or stakeholder is assigned responsibility for or "ownership" of a risk. They are not going to do anything about the risk unless their boss beats them about the head and shoulders.

The person assigned to handle a risk should be someone who can affect the outcome of the risk. Who is going to contact purchasing to ensure that the equipment arrives on time? Not the director but someone on the director's staff. Assign the task to the person who will take the action. Who will contract with a cloud vendor to host the application? Probably purchasing and contracts, but those are departments. You need to name the action officers, the actual people on the hook. Give them a schedule, give them a priority, and help them manage their workload so that your items are completed when you need them. It's good to also include the senior managers so someone with a big hammer can swat problems away, but don't list that senior manager as the person assigned to the risk. Also, you don't really want the senior manager, who probably has only the foggiest understanding of the risk and associated actions, going around solving it. That adds a new risk that he or she will either take up your entire day with inane phone calls and emails or mess it up completely. When you're in the project meeting and you decide that someone in purchasing is needed to help reduce a risk, ask the purchasing director who has been assigned to the task. Don't let them say "Me"; that's as insane as assigning every programming task to the project manager.

Opportunity Management (NOT)

Much is made about opportunity management, which is the plus side of possible events that could be included in the project and make the world a better place. This is a crock intended to make us feel good in management meetings. "Opportunities" almost always involve increasing the scope of the project or adding dependencies on other efforts and therefore are not sugar-coated wonderfulnesses. I'm not a huge fan of including them in the risk register. They should go into the requirements list or product backlog as low-priority items. In summary, there is no opportunity management.

Requirements Management

If there is anything that the software world agrees on (and I'm not sure there is), it's that requirements are key to the success of a project. I've read numerous papers on the subject. They all say that without a robust and complete set of requirements, a project will fail. I think we'd agree that if you have the wrong ones you are destined to fail but that even the right requirements are right only for a while. I've seen many slides that showed the correlation between high rates of requirement change and failed projects. The old school of thinking, which sadly is still prevalent today, is that you should send out business analysts to talk to the business users to hear what they need from the system. Most IT shops have a business liaison function that is often filled with former testers or developers who didn't like coding. The business analysts write down what they heard in what they perceive to be objective statements of required functions and review that with the users, who go back and forth with them and eventually approve a document filled with these statements.

So right off the bat a transfer and translation has gotten into these statements. The analyst tried to bridge some of the gap between the business users and the developers, and they've abstracted some of the functionality into higher-level statements that may or may not be truly understood by the business users. Business users don't want to appear stupid, so often they assume that the analyst's phrasing captured their needs even if they aren't comfortable with the wording.

Sometimes the users aren't good with abstraction at all. I had a client who needed to have various forms electronically signed by multiple people as things were processed through the system. I wrote a business function as "The system must support electronic signatures throughout the workflow. A document can require more than one signature at each step in the process." The client didn't get this at all. "No, see the purchasing manager needs to sign it, and then we need another requirement for when the sales manager approves it, and then…" There's no reason why business users would know that once you implement the ability to handle electronic signatures, you can use it everywhere. To their minds, the requirements listed a fuzzy, vague statement about signatures and completely missed the point. The business analyst told the users that this statement covers all signatures, and eventually they accept that or just stop arguing for fear of appearing dense. They don't see how that statement could possibly cover all the signatures and steps in the process. The message they get is that that nothing in the requirements is expected to be clear and that vague terms actually are the specific terms they would look for, so they stop critiquing the requirements. Things now get missed. Mind you, all of this missing is happening between two people who don't speak computer. We haven't even involved the developers yet. I think you can imagine how wild the miscommunication can be between an engineer and the end user.

One school of thought maintains that changes in requirements are a problem. They argue, on the one hand, that if requirements are changing wildly, massive amounts of rework will have to be done and budgets and schedules will be blown. Imagine planning to build a house, and when you finish pouring the foundation, you are told that the house is supposed to be ten feet farther to the left. On the other hand, the notion that one can document all aspects of a system in sufficient detail to be able to then build the system with only minimal end-user involvement ignores the real world. Requirements for real world systems are going to evolve, and that's just a fact.

Over the years I've seen a number of approaches for managing requirements come and go. We've tried documents that had all manner of statements that began "The system shall..." or "the system must..." These were always a blend of design decisions, problem statements, personal desires, pseudo code, and process statements. We'd go off for many months building a system based on this approved document and fabulously disappoint our users.

Various modeling techniques have tried to show the current and future systems in a way the users would understand, but most failed—badly. One of the failure points here is that a very good user of the current system might not be able to express the business need in any manner that isn't just a recasting of the existing world in new words. You bring a person into a room who knows everything there is to know about how the enterprise does things today and in essence ask, "What would a system have to do to support a completely new way of doing business?" They might actually be the worst possible person to bring into the room. They achieved this position by being very good at the old way of doing things, and you're asking them, "What do we need to do to make all your skills and talents useless?" You are likely to get requirements that are somewhere between what the actual need is and today's way of doing business. So requirements will change down the road as the business changes.

Old school is to have a formal requirements change review process. This is almost always led by the IT department because the business users have no interest in it. They've said something needs to change, and that should be the end of it. They do own the business, after all, and why they have to sit in a room with a slew of engineers to justify changing *their* system is beyond them. The change control board or CCB will have an agenda. They receive the request, have someone talk to the requestor to better understand the request, have someone estimate the complexity and impact on the system, and have someone estimate the level of effort for the task. In a "properly" run CCB, these steps will all have been taken before the monthly scheduled meeting. Now they sit in the room and go through the ritual of approving the request. This is a colossal waste of time and money. There are few people in the room who can actually approve the budget. There are also very few requests that have low value but would require massive resources to execute, and those can be weeded out right away. In fact any change that requires massive resources should probably be looked at as a completely new project in the first place. So we're almost always looking at requests that have medium value and require a low-to-moderate level of resources. The users just want the new feature. They resent having to attend the meeting, which they perceive as IT bureaucracy, and the IT department resents the user's lack of appreciation of how organized they are to allow the user to optimize their precious resource...the IT department.

The reason I believe that the Agile approaches are gaining such traction is that they eliminate the need to do the following:

- Know everything about a system before you begin

- Create a language that is foreign to everyone on the team

- Prevent and resist change

I am also a fan of the User Story approach in which the need is expressed as a task the user must perform. For example, "As an order entry clerk, I need to be able to look up the price for the item the user is requesting by manufacturer, color, or category." It has no computer-ese in the phrasing. It's a business need. All of the lower-level details will be worked out with the user in the room. The developers and the user will talk to each other and develop their own lexicon for talking about tasks like this. They might start calling products "things" or "items" or whatever they work out together. It won't be some word chosen by a well-meaning but now uninvolved analyst. As they work through the details, the developers and the users will discover together that they also need to be able to look up the price by quantity. But in the Agile world, there's no harm, no foul in this "late" discovery. I have other thoughts on the plusses and minuses of Agile, which can be found in a later chapter. For now, rest assured that although I believe there are some good features to it, this skeptic does not believe Agile will solve all problems.

Thoughts on Agile Development

Agile is a methodology that is getting a lot of attention of late. Many organizations believe that Agile will solve all the problems of software development. Agile does solve some, maybe even many, but it is not a cure-all for all methodological ills. This is not a book about Agile's pros and cons, but I will cover a couple of things that relate to our friends the engineers and the supporting characters on a project. For the record, I am a huge fan of Agile. I think we need to be moving toward this approach, but we need to do so with our eyes open. It doesn't come free and if we turn it over to the process people we'll be right where we are today with a rigid cookbook of steps and wondering why we're not making progress.

Requirements

Agile is where the User Story concept comes from—stating the need in terms of a user doing a task—and why it is easier for users to understand. It's in their language. Look at the difference between a traditional requirements statement and a user story.

Traditional Requirements Statement

1.0 The system shall support the entry of product orders.

1.1 The system shall support the entry of multiple lines of products for each order.

1.1.1 For each product, the system shall provide a five-character alphanumeric field for the product ID.

1.1.2 For each product, the system shall provide an integer numeric field (max 99,999) for the quantity of product ordered.

User Story

As an order entry clerk, I need to be able to record the items a customer wants and the quantities of those items so that the system can calculate a price, check the stock, and estimate a ship date.

So Agile has it hands down on requirements. Everyone can understand the User Story, but to an order entry clerk, the detailed enumeration is confusing and does not help anyone spot the missing data items.

The refined, detailed requirements come out while the system is being collaboratively built with the order entry clerks, who are some of your business users. It evolves naturally as part of the process.

This continuous refinement of the requirements in Agile can be unsettling to engineers who are not used to working in an Agile fashion. They are still trying to write the most perfect code, and the ground is constantly shifting under their feet. As their manager, you have to reassure them continually that perfection is not what we are striving for in the early phases of the project. We're focusing on sufficiency and function completion. We want to get something out there so the users can better see what it is they've asked for. The goal is to get them something very quickly, even if it's ugly, so they can better state what it is they *really* need. We, the engineering team, are going to make sacrifices now so the users can better describe their needs.

Working this way is genuinely hard for perfectionists. They have to change their development focus from perfect, or at least very good, to quick and working. It is a tough transition. You have to spend a lot of time walking the floor to make sure your guys are not writing perfect code or trying to do things "the right" way. Agile says that good enough is good enough, but that's not good enough for an engineer.

You need to praise them each time they rough out a function. Let them know you know it's not of a quality they would usually create and thank them for making this trade-off. Focus on how happy the users are to see the function and how much happier they will be when you finally get to refactor the code into something more robust or "proper." They are fighting every instinct they have to do things in this "sloppy," "hack," or "inefficient" mode, and it truly is a heroic effort on their part. Make sure they know you know what they're going through.

Design

On the bright side, engineers love the sufficiency concept of Agile design. In Agile you do a design only to the point that everyone knows what's being built. The danger here is that designing this way can be a license to not design at all and just go straight into coding. You have to make sure that there are no loose ends or interfaces that require collective thought and documentation. You have to make sure that the services have been discussed to the point that no two developers will be out there writing duplicative code. Designs in Agile can be nothing more than a bunch of scribbles on a white board or emails back and forth.

Although the form of the documentation is not important to the engineering team, others will need to review the design at some point, so you are never fully unburdened from documentation. The enterprise architects of an organization hate Agile's design as it gives them nothing robust and formal to review and approve. They will feel like the team is completely undermining their role. To address this concern, you should invite them to the design sessions at the beginning of the sprint so that they can see what's happening and have the opportunity to chime in with their thoughts and needs for architectural compliance. In some cases, they may even be so motivated to make up the actual drawings and documentation for you.

Testing

Testing is different in Agile than in traditional development
methodology in that it is continuous. In non-Agile schedules, there's a
testing phase at the end of the development phase, and that's where
the testers spend a large part of their time. In Agile the testers are
there all day every day. They are testing as the developers are coding.
It is much more interactive, and the communication between the
engineers and the testers is constant. As the project manager, you have
to ensure that they are playing nice. Many testers are extroverts, and
they can overwhelm the energy needs of the introverted engineers.
You will have to coach the testers on the needs of engineers and guide
them to a happy medium between the tester's information needs and
the no-people-time needs of the engineers.

The Supporting Cast

Almost all of the processes and training of the supporting cast for IT
shops is geared toward large drops of code that happen infrequently.
Agile messes this up in a big way. Almost all of the supporting cast
will hate this. I have seen shops in which the software development
team is operating in an Agile fashion but no one else is. This causes
some obvious problems.

The Unix system administrators are always grumpy, and your proposal
to build a system that requires unknown resources is not going to
make them any happier. You don't know how big the database will be,
how many transactions there will be, how much the processing load
will be. However, I argue that you *do* know a rough order of
magnitude of the problem. I would be generous on the growth
estimates. Then confidently assert your ROM requirements as if they
were more firm than they are. That lets the Unix system
administrators get to work on your environment without undue worry.
If you go in low and constantly inch up on resource requirements,
they will wait to jump you in dark alleys. They like a solid, stable,
knowable set of code that they can then optimize or at least contain to
a known set of resources.

The Database Administrators (DBA)s also have this angst. They want a complete specification so that they can build their perfected tables, but you aren't able to tell them everything they want to know. They will have to recreate the table structures frequently. They will hate this. As with the other resources, you need to ask for a large amount of database space even if you can't justify it by the few tables you will install initially.

The security folks have no good way of dealing with Agile. All of their processes are set up to scan finished, stable, known products with their vulnerability testers; review the final documentation; and inspect the code if necessary. It is very labor intensive and usually takes a couple of days at the least. Their method is very batch oriented. It assumes a quarterly, at most, but more likely a biannual code release, with a full week to review it and give their blessing or respond with a list of problems that you'll fix, they'll rescan, and so on. They are in no way prepared for a weekly code drop. My recommendation is to be proactive on this front. If you tell the Information Security (IS) team that you're going to do weekly code drops, what they'll hear is, "I will eat up all your resources for the rest of time," and they will stop you dead in your tracks. Instead, find out if there is an architecture that they are comfortable with or that they can get comfortable with and propose scanning the entire first drop as usual. From that point forward, you will provide them with all deviations from this approved architecture, and only if the deviation is of sufficient magnitude (to be determined by the IS team) will another full inspection, scan, or review be necessary. For most systems this should be acceptable.

Your trainers are also going to hate Agile. Like the IS folks, they will look at Agile and run for the hills since you're going to be changing the user guides, training materials, and everything else under the sun on a weekly basis. They can't support you. The trainers have to get on board with the concept that the system is going to have small incremental changes that are requested and designed by the business users so they, the business users, will be training each other on the incremental changes. The training department just has to get them jump started at the beginning and then do updates as the number of changes merits. They can decide when that point is reached. Watch for them to start breathing again, and then you know they've heard you.

Management

Many organizations look at Agile and are captivated by the notion of deploying operational software in less than a month. It's never happened before. They see it as the shining star, the golden goose, the Holy Grail. What they fail to appreciate is that you don't get a complete system in a month. You get portions of a system early and it continues to get better, but it still might take a year or more to get to the final version. What they also fail to grasp is that you cannot tell them how much time it will take or when exactly they will get to that final version. This is completely lost on them because of the shininess of the "systems in a month" message.

If you are running an Agile project, you must stress this constantly to management. They have to know that you are never promising that the final system will be ready on such a date and at such a cost. You can stop at that date or that budget value, but you don't know if you'll be done.

I've heard lots of folks say Agile is designed to fail quickly. I hate that turn of phrase, but the concept is valid. By building small portions of a system and getting them into the hands of production users, you can very quickly see if you have missed the boat. It sure beats spending eighteen months building a system only to find out that the users were wrong on what they needed. But failing early is not a goal. The risk of building the wrong system can be mitigated with Agile, but it never truly goes away.

The majority of changes to an application happen in the first year after deployment. The evolving requirements, managed in the User Story Backlog of Agile, incorporates those changes into the development cycle. In one way this is wonderful because you are getting to the *actual* requirements much sooner. However, what does this mean to the overall schedule? Compared with a traditional approach, you're building the system and the first 60% of the changes into the project. This might require 50% more resources than if you had only built what requirements users could state at the beginning of the project. So in some ways Agile projects could appear to take *more* resources than traditional approaches. This is another message you need to reiterate to management.

Software Reuse

Software reuse, like Agile, is another one of those shining cities on the hill. If we just reuse software then all our costs come down. We're maintaining less code; we are optimizing our investments; life is just great. On paper that's precisely the case. In the real world software reuse hardly ever happens. Why is this? How come we can't get people to just reuse a chunk of code or a service that does this already? There are two main reasons.

It Doesn't Do This Already

When a developer writes a piece of code, it's to meet some specific set of requirements. An astute engineer might see ways to make it more generalizable, but the engineer could not be expected to perceive every possible future need that the code might have. It might do much of what is needed by the new system, but it might not do it all or not do it using the business logic of the new system.

So now a developer and his or her project manager have a choice: modify the existing software to also meet their project's needs or write some of their own. Writing their own means they can optimize the code for their system. It will be clean, quick, and in the same style as the rest of the code being written. Modifying the existing code requires coordinating with another project, perhaps with a configuration management team. There will be meetings and design reviews and arguments and discussions. It will be more involved to test because the original system *and* the new system will have to be exercised. More approvals and reviews will be needed. Enterprise data architects might get involved to make sure your using the approved data structures. The security staff will be very concerned about cross-system data leaks. They might delay deployment of this modified piece of code, which would delay the new project's deployment.

More often than not—*much* more often than not—they choose the easier path and write their own code. In fact they can probably get it written, tested, and even deployed before anyone notices that they've replicated the functionality. At that point it's very hard to justify pulling it out and modifying the code.

Software Development Is a Creative Process

Software development is not a routine process. A great deal of creativity is involved. Each piece of code is like a new painting. Yes, it may be a painting of a person on a bench, but give that assignment to two different painters and you will get two very different results. Developers have a style; even with coding standards, there is a style to the code. It reflects their problem-solving process. Think about how people do crossword puzzles. Some go for all the clues across then down. Some try to solve blocks of the puzzle at a time. Others will try to solve for all the short words first and then build up to the longest words. Software development also reflects this kind of personalized problem-solving approach. So when a piece of code might be out there that closely meets a new system's needs, it is very unlikely that it was written in the style of the current developer.

To the developer, this is a cognitive problem. It's akin to being told to do a watercolor portrait of a woman on the subway but start with the Mona Lisa's face from da Vinci's original. It can be done but the painter will struggle throughout. To the engineer, it would be less stressful to simply paint a new woman's face into the portrait. It will remain in their style. They won't have to adjust the painting if it changes from watercolor to oil; they won't have to change their brush strokes. Reusing code is just plain mentally harder to do, so if engineers can write their own before anyone notices, they will almost always do so.

There are other reasons why software reuse has not saved the day. It can be hard to find the already existing function in the enterprise of code. It might not be documented in a way that makes it easy to find. It might also be hard to figure out how to use it. It might be in another programming language, unknown to the current developer.

For these reasons the benefits of software reuse have not been realized, and you should not bet the farm on your enterprise seeing it happen anytime soon.

Design

Engineers either love the design phase or hate it with a passion. Those who love it are the architects and optimizers. It's the phase in which they get to dream and scheme about ways to implement the system. They imagine data flowing back and forth through the layers—secure, properly bundled, and efficient. They are in hog heaven. Meanwhile, those who hate it are in a serious funk. They want to get on with the coding and think that all this stewing about with architecture diagrams and circles and arrows is just an academic exercise. They view the designers as fussbudgets. Nothing gets done until coding is finished, and the fussbudget's design phase is just delaying that. They resent the designers, and the designers view their opposites as code monkeys unworthy of appreciating the genius inherent in their design—as cowboys who will only mess things up if left to their own devices. To a degree they are both very wrong and very right.

Software design has changed dramatically over the years. Back in the day a system was broken into very large components,

> Many trees died to bring us this design -- Mon Mothma, before the Battle of Endor (okay, maybe not)

and each of those major components would have a top-level design. It included the overall notion of how the system would work. The details, surprisingly enough, were then figured out in the detailed design document, which truly *was* detailed. It went all the way down to something called pseudo-code, which was almost compilable. A ritualistic, detailed design review would take place during which, in theory, someone went through all of the documentation and spotted the errors: a missing variable not being passed in, a logic loop that missed boundary conditions. We'd revise the document and move ahead.

When a design was done to this level of specificity, it got large. I recall one system for which we delivered the detailed design in two five-inch binders with double-sided pages. No one on this earth will read all of it. To make the review useful and to spot the inconsistencies, reviewers not only had to read the design, but also keep it all in their heads. In addition, they had to have read the top-level design (which was about three inches of paper) and remember all of that. The reviews proved to be a colossal waste of time and Canadian old-growth forests. We'd then go on and build the code, and when we were done, we'd do the complete "As Built" documentation suite, which was the detailed design documentation completely redone now that we had found all the flaws in the higher-level logic of the original detailed design. Mind you, this was for systems that were far less interactive and complex than today's systems.

Today there are foundational layers available that had to be hand-crafted in the past. Their availability has had a profound impact on software design. Once one chooses the architecture for an application system, much of what used to be a top-level design can now be assumed. This supports the engineers' argument that design is a waste of time. "Okay, we're in a LAMP (Linux, Apache, MySQL, PHP) stack. Let's get on with it already."

But there are still some rules, conventions, and standards that should be agreed to before one lets the team go hog wild with the code. An obvious one is what to do when errors occur. Will there be a central error log, or will each layer maintain one? Will the user be notified, and if so how? Do you store detailed results in addition to summarized results or always calculate based on raw data? Also, and I hesitate to say this, you need to establish some basic coding standards. At a very minimum, you should agree on whether variable names indicate type or purpose or are just descriptive, how you note the beginning and end of logic blocks (e.g., loops, conditionals), and what your documentation blocks must include.

However, the architect/designers are also right. Someone needs to spend time identifying key transformations to the data and where they should happen. You can have holy wars on what methods should be contained in each object and what logic should live outside of the object itself. I recommend leaving these decisions to the designers. They are the ones who most treasure neatness and simplicity. They are the ones who look for clarity of purpose and goal for each function required in the system. Often the engineers hate this. They lose control if the higher level manipulations are not in their objects. They also might not be able to complete some of their work until these other functions are available, and this makes them feel vulnerable and helpless.

As the project manager, you need to be very visible and vocal during the design phase. There will be fights and disagreements, and, as I covered in "Dealing with Fights on the Team," you need to step in when appropriate. You also have to make clear what your needs are out of the design phase. You need to say out loud that you expect to have the coding standards described and, more important, adhered to later on. You need to say that the goal here is to reduce rework and make sure that what we have hangs together when all the parts are complete. It's to establish the philosophy and style of the code to be developed as much as it is to ensure that everyone understands how it's all going to go together.

User Interface Designers

Graphical User Interface (GUI) designers are a fascinating contradiction. Frankly, I have really liked every one I've worked with but have wanted to poke my eyes out for working with them all at the same time. Why? Think about the skills required to be a good interface designer. Think about the problem statement itself. You have to take a real-world problem and process—something that has nuance, interdependencies, required information, vague information, approvals, processes, requires searching for information, and the like—and translate it to the most simple-minded object in the world (a computer) and while doing that never lose your real-world train of thought. It's a rare gift. You have to understand the mental processes of the persons sitting in front of the computer. What words do they use? In what order do they want to do things? What's the way they'll want to look for things they need to complete the task?

Developers all believe they are good at this: brilliant, gifted in fact. I sure did. However, very few of them are any good at all. In fact, most are horrible. Having now worked with some gifted user interface designers, I can say with authority that I was weak on my good days. It's one of the reasons why the business users believe that engineers are idiots. Engineers create these unusable systems that are structured to make the code more efficient, not make the business task simpler or more efficient, or at least that's how the business users interpret it.

The other curious thing about user interface designers is that they are completely dominated by their methodology. Nothing you say will change this. It would be like asking Picasso to change his painting technique. You might be able to make him do it, but would you get anything good out of trying?

These designers have agile minds in that they must constantly change business domains. They have to get completely into the skulls of the users, see the world through their eyes. They have to know a fair amount about how computers work and think, too, about how systems access data, how they hold information, how they maintain states of things.

They then have to take the "Lego blocks" of the computer world and build an interface.

Users often find the designers wonderful at first. They've had to put up with the requirements analysts for months, then the inane questions from the technical team, plus they might have had a slew of questionnaires from the change management people. No one seems to understand them. Then a wonderful thing happens: someone comes along who speaks their language, who asks them what *they* like— colors, styles, placement of things. This new person seems to understand what they do and shows them something real! No more paper, no more pages and pages of words that are abstract techie decompositions of *our* business that are pretty hard to decipher.

Designers come up with the wireframes—*finally* something that looks like the system users have been so excited about. That they've been waiting for, for *ages*.

One problem can be that the conversations with the users, if left unsupervised, can lead to outrageous scope creep. You might have a system targeted at human resources hiring, but once the GUI designers start talking to HR, you have features and functions on the screen that cover annual performance review, annual EEOC compliance reporting—you name it. It's absolutely true that the data relate to one another. It's extremely likely that incorporating the features into the GUI now would be so much easier than doing it later, *but* it sets expectations for the user community that they're getting those functions and it scares your sponsors that you're trying to grow your contract/effort, that you can't be trusted, yada yada. So you still need to keep the developers away from the users, because they'll annoy each other. You *must* have someone on the analysis team or perhaps you yourself in the room.

It's not that the GUI designers are evil; it's that they need to get into the skulls of the users, and those skulls are covered in functions that go beyond the project mandate.

Your GUI designers are also your best change agents. They are your face to the users, the first face they can understand. You *have* to use this opportunity. It is your best chance to get the users to adopt the system. They will now have skin in the game. It's their baby; they own the system. Why? Because they designed it!

Estimating

Entire volumes have been written about estimating software projects, so I'm not going to try to do a literature review in this section. Instead I'll try to share what I've had the best luck with and some of the shenanigans that go on during the estimation period of a project.

Estimating is hard. Lots of very smart people get it wrong. People with decades of experience get

> We view IT as repeatable and technical, when in fact it's creative. You wouldn't ask da Vinci "how many brushstrokes will it take you?"
> – from 2013 AgilePalooza Conference

it wrong. They get it wrong every time. We ask them to help estimate the next project because they get it less wrong than other people, but it is nonetheless wrong. How can that be? Customers assume estimators are a bunch of charlatans trying to bilk their company out of additional fees. Management thinks their inaccuracy is simply incompetence. It's actually neither of those things (at least among reputable people).

All estimating is biased and subjective. That might seem like a pretty strong statement, but allow me to explain. Before one can estimate something, someone has to describe it. The person providing the description is doing so in a certain frame of mind. A description is necessarily bound by what the author assumes is affordable. The author is probably concerned that the project will cost too much, and if it costs too much, then management won't approve the budget. So without trying to be deceitful, the person is likely to describe the system as simpler than it might actually be. He or she might consciously or unconsciously make assumptions that would reduce the level of effort. To get management approval, certain things the system must do might be phrased in very simple terms that are correct from a business perspective but that downplay the technical complexity involved. In doing so, the writer may be unaware that a landmine has been placed in the project from the beginning.

Often the description is written by someone with very detailed knowledge of the business. Sometimes this deep experience in the area can lead the person to skip over things assumed to be common knowledge. The writer may forget to describe the exceptional cases or an annual process that "everyone just knows about"—except the person who has to estimate it.

Writing a problem statement for a new system is not a normal task for a business unit; it falls into the special projects or "other duties as assigned" category of tasks. The person who takes it on already has a day job, and this is an extra task, so either they were dragooned by management or they volunteered. Why would they volunteer? They might be trying to curry favor with their managers; they might be truly interested in the project; or they might want to get some variety in the work day. Regardless, they are not guaranteed to represent the views of the user population who will be approving requirements, validating designs, and otherwise approving the actual work to be done. So we start with a problem statement that is consciously or unconsciously biased to produce a lower estimate.

The estimate is first examined by a technical manager, who is personally motivated to do a good job and come through for the company (companies often have incentive plans that reinforce this behavior). These technical managers look at the problem statement through their own shade of rose-colored glasses. They too gloss over the complexities and accept as reasonable any stated simplifying assumptions, because they want to get on with the response phase of the proposal. By personality, they are usually doers.

Technical managers will likely draw an analogy between systems they have built in the past and the proposed problem description. The technical manager will select the analogous project based on an assumed technical architecture. "This job is like that Acme job we did two years ago, and that took a team of six people eight months to do." But is it like the Acme job? Maybe a little. What drove the Acme job to require six people eight months to do? Are those the same drivers we have in this work? Maybe some are, but probably not all. Was there anything special about that team of six people? Are they the same six who will do this work? Probably not all of them, and perhaps none of them will be involved. Were those six even representative of the team you'll have? There are no guarantees. For that matter, is the manager remembering the Acme project correctly? Were there really only six people on it? Did that include the whole team? Was it eight months or ten? As you can imagine, choosing Acme as the model colors the process with the manager's subjective experiences and a number of assumptions, many of which could be false.

Now with only two hops we've layered the technical manager's set of biases, recollections, and experiences onto the original biased and assumption-laden description document. To make it better, we've even added an assumption on the technical architecture!

In some cases the technical manager is the sole author of the estimate, but usually the document is taken by the technical manager to some subset of trusted engineers. Without thinking about it, the technical manager almost always begins the conversation with, "I have a new piece of work that is a lot like the Acme job we did two years ago. I need you guys to look this over and come up with an estimate." Boom! We've biased the reading of the problem statement with the polarized glasses of the technical manager's Acme project. When the team reads various sections that might be ambiguous, they are seeing them through Acme project filters. They resolve into the Acme project mold, and sure enough an estimate comes back looking a lot like the Acme project.

That's the simplest case. Let's now take a look at a competitive proposal. We start with the same biased problem statement. It will arrive at a contractor's doorstep through the contracts office and then go to one or more business development leads. Business developers are the sales persons at a contracting company. They are the ones who do the face-to-face work of getting the company known by potential customers so that they'll have a shot at bidding the work. If customers don't know your company, they won't ask you to bid that work, so these folks have a relationship of some sort with the customer. In some cases they might even have had conversations with the customer that helped shape the problem description. Either way, the business developers are viewed as the subject matter experts within the contractor's operation. They are the ones speaking authoritatively about the requirements of the project and the business to be supported, and they have a key role in developing the strategy for the proposal response.

Business developers are not evil people, but like everyone else they have their own motivations and incentives that drive certain behaviors. Management wants the business developer to bring in lots of business. To inspire that result, business developers are usually given a middling base salary and then a bonus or percentage of every piece of work that the company wins, so with more wins, they get more money. All things being equal, you will not win a job if you come in with the high bid. You won't win every time by having the low bid either, but two equally rated bids will be given to the lower cost option, so the business developer wants to bid the work at a cost that will maximize the probability of winning the work.

With this guideline running through at least the back of their minds, business developers bring the work to the appropriate internal business unit that will execute the work. The business developer has a figure in mind that will win the job. In discussions with the technical manager, the business developer will be driven to talk about the work in terms of some other project that came in around that figure. This comparison is passed on to the technical manager, who layers on his or her own experiences, and we rejoin the loop above. So we add at least one other layer of biases, subjective experiences, and the like.

There are formal estimation methodologies out there. They can include all manner of steps and reviews and documentation and double checks. But they are all based on the problem description and the polarization of the multiple layers of interpretations that have happened upstream. Even if it was an accurate interpretation of the problem statement, as I pointed out earlier, the problem statement itself might not represent the work that would have to be done to succeed in the project. Finally, all of the above is based on the assumption that the goals and details of the project will not change during the period of performance for the project, and in the real world that rarely happens.

Well, that certainly was a depressing little journey! So what now? We still have to give someone an estimate. We still need to respond to the proposal. If the very premise of the description of the work to estimate is flawed and it goes through a series of filters and interpretations to derive an estimate that will be used to judge the success or failure of a project, shouldn't all estimates be failures? Yes, and if we stayed solely with the requirements at the beginning of the project, they would be. A good estimate is someone's best guess and nothing more. At some level the estimate is subjective, biased, and most likely wrong by some double-digit percentage. Here is my next blasphemy: we need to accept this.

Estimating is a human activity, and there will always be error in every step of the process. If we want to stay within a fixed budget, we will have to trade off functionality. If we want to fix the time required, we might have to add people (and with them cost) or again, trade off functionality. If we want to do everything on the wish list, then it might cost more time and money. There's just no way around it. We need to accept estimates for what they are: good, educated guesses.

The fact is, whether or not we want to admit it, every project adjusts functionality to fit the amount of money and time allocated. From day one of a project, functionality is being negotiated downward to fit the hours allotted to the project. Agile methodologies do this explicitly. If you're following any other approach, you're doing it on the fly but you're still doing it.

The explicit, up-front tradeoff on functionality is one of the reasons people are so excited about Agile development methodologies. Look! The users decide what is most important; we're doing small duration efforts, and we're getting what we expected! We've dispensed with the broad problem statement and the large single estimate. Instead we've created a set of smaller problem statements called User Stories, and we're engaging the end users to keep them on target. We're accepting generalized estimates for each problem statement, and we make sure the individual sprints are small enough that we can never get too far off target or in too much trouble.

I have other thoughts on Agile, but before we all jump up and down with joy that we've solved the problem, let me throw a wet towel on this celebration. Agile does no better at providing an estimate for the full-up total project at the beginning of the project than any other process. The difference is that no one notices until they've used up all the funding but still have user stories to complete. Likewise, Agile won't be any better at telling you when you'll be done than any other methodology. But I'm getting off topic. This section is not about the pros and cons of Agile; it's about estimating.

Despite what I've said so far, estimation is not a pointless ritual. There is either an internal agreement or an actual contract that will be established to approve and fund your project, and it's going to be built on your estimate. It's something to take seriously and to try to do well. But how does one develop an estimate? Over the years many approaches have drifted into and out of favor.

Source Lines of Code

Just after the days of knots-in-string computing, we used to estimate projects by the number of lines of code needed. We'd estimate the source lines of code or SLOC. It was agreed by the powers that be, at their annual Powers That Be Jamboree, that your average coder (what we called software engineers in those days) could produce 30 lines of code in a standard work day. If you figured out the number of lines of code, then voila! You had the total number of coder days and life was good. But then…how did you come up with the total lines of code? Well, by analogy. We'd think back to a project we did in the past, look at that project's code repository, and count the lines of code. We'd then apply a certain amount of "Kentucky windage" if we found our developers were getting cleverer in what they could accomplish in a single line of code and to account for either more or less complexity than we thought was in the project being estimated and—ta-da! This project will take 10,300 lines of code!

Lots of people worked on coming up with standard metrics for common functions. A write to a data file in COBOL was 75 lines of code, or putting up a menu and getting the user's choice was 200 lines of PL/1 code. There was a lot of diligence and detail that people put into these metrics to try to take more of the guesswork out of estimating. They'd survey and count out as many code repositories as possible to try to get the number more and more accurate. It worked…poorly. It assumed that we could accurately enumerate each read, write, menu option, and report (for that's about all most programs could do back then) at the start of a project, which I think I've shown you can't do reliably, but it was all we had at the time.

People built large, elaborate SLOC estimating models. The constructive cost model, or COCOMO, is perhaps the best known. You'd enter the number of screens, inputs, outputs, data fields, reports, team size, experience, you name it, and out came a number— poof! Your estimate on a plate. People went to classes to learn how to use COCOMO. They'd then ask the engineering team more questions than you could shake a stick at, and the model would perform a regression analysis on your inputs and the data from a slew of historical projects. Over time the model would learn how you did and, in theory, give you better future estimates based on the productivity of your specific team—that is, if you had a phase in your project to populate COCOMO with all this data, which almost no project does. So models like these produced similarly formatted but no more accurate estimates than the old manual methods. SLOC was not a reliable indicator of project completion. Because each coder had a unique style and way of doing things, the variability of the SLOC estimate was enormous. One could not say with any assurance that if the project was estimated to take 10,000 lines of code and 4,000 were in the repository, then one was 40% done with the project. You just could not use the counts that way.

Counting lines of code was, however, good for starting religious wars among developers. If your team could produce two times the number of SLOC in a day, then you were two times better than the other guys. Huge arguments would ensue about what was and was not a countable line of code. Did variable declarations count? No, that's not executable. Okay, what if I declared my variable within the line that it is first used? Does that count? Sometimes people would split out a function across three lines of code when it could easily be done in one. Is that three lines or one line? Folks wanted a better way.

Function Points

Function points was the next popular approach of the past that tried to remove some of the subjective judgments from the estimation process. To use function points as an estimation tool, one would first break up the system to be estimated into five categories of business functions: inputs the system need to do the job, outputs it must produce, number of files it uses, searches it must perform, and external systems it must interface with. You'd then look at each one of those and assign it some number of function points based on the complexity of the item and then count up the total function points. The points would, once again, be based on your historical data, and somewhere you also had a conversion factor for function points per programmer (what we called software engineers by this time) per day.

But function point analysis assumed you could name all those items, analyze them with some degree of certainty to assign function points to each, and then sum them up. You had to know how many database tables or files (depending on how far back you want to go) you would have up front. You had to know the number of screens on which a user would be making selections, the number of reports and what set of searches was needed to populate that report, the number of other systems that would need the data from your system, and the searches to support that. You also had to apply Kentucky windage to decide how complex something was and how many additional function points that complexity merited. Function point analysis did not actually change much in term of accuracy, though it did get us away from counting lines of code, which while fun for the OCD-types, is not much fun for anyone else.

Management Loves Models

In this case, I don't mean fashion models but estimation models. There's a legitimacy that the perceived (albeit false) rigor and objectivity give to a fancy estimation model. I was once working a proposal for a medium-sized application for a state agency. They wanted to perform the work on a fixed-price contract basis. This makes management cringe because a fixed-price contract places all the risk on the contractor. No matter what happens, and usually no matter whose fault it is, the price stays the same. You can make a good profit if everything goes great, and you can lose your shirt if things go poorly. So management was very anxious about the estimation of this one. They asked three of us "gray beards" (old, experienced types) to come up with an estimate independently. We all went home that night, worked on it without sharing any thoughts or assumptions, and came back the next day with numbers that were within 2% of one another. We got a good chuckle about it. Management would be thrilled that their three trusted advisers (or so we assumed we were) had nearly identical results. Surely with three data points and such a small standard deviation we know what the number is. Let's march forward! Tally ho! But, um, no.

In fact they hated it. It scared them to death that there was no spread in the estimates for this highly risky (financially) project. Rather than assume that we clearly understood the problem (because three people working separately who came from very different parts of the company and had very different experience bases had the same estimate), they assumed that we had no idea what we were doing. So, they brought in the COCOMO model and the modeling team—those brave individuals who went to the class on how to input numbers. They spent a week entering numbers, counting things, and tweaking the dials on the model. When the answer came in…it was smack in the middle of our 2% spread of three hand-done estimates. But this one had rigor! Management loved it. They praised the modeling team. They said nothing to the gray beards.

But Management Doesn't Trust Models

Now, as a side note, we moved ahead with this proposal. We three graybeards wrote up the technical approach and management sections of the proposal, which allowed us to watch in horror what happened later. At the next-to-last review of the proposal, a very senior manager who had not yet been involved in the proposal was present. He was very worried about the fixed-price nature of the work. He asked how we got to this figure. The proposal team leader proudly stated that they had used the COCOMO approach. He described how the fully trained team had gone about entering in their numbers and working through the problem and how they were fully confident in the result. The senior manager said, "I don't trust COCOMO; who else has looked at this?" The team leader replied that the three of us had come up with independent estimates that were nearly identical to the models numbers. He liked that no more than the previous managers had. "You can't do this work for that amount, it has to be twice that," and so it was written, so it was done. We bid the job at twice the price that was estimated by three humans and one model. We lost that proposal. The winning bid was exactly in the middle of the gray-beard estimates.

And Then Came…

After function points there followed all sorts of variations on a theme. PERT, SLIM, Use Case Points (notice how similar that sounds). Some of today's Agile approaches use a point system as well. The team looks at a user story and decides how many (1, 2, 3, 5, 8, 13) points it would take. The idea is that you estimate how much bigger each task is relative to an arbitrary base small task with a value of 1 point. A modified Fibonacci series of values is used because…well, why not? They initially estimate how many points they can do in a day (their velocity), and over time their velocity estimate gets more accurate.

What all of these approaches just can't get past is that we are *always* estimating future work based on our ability to find a historic analogy for it. At a very basic level, we are always doing estimation by analogy.

It should come as no surprise, then, that the best estimators are those who have the most accurate analogy. Who are they? Either the person with the greatest depth of experience in the target technology or, more commonly, your most experienced engineers. In this case, the experience that counts is number of projects and problem domains, not longevity. An engineer who's been working on maintaining a core business system for the past ten years is absolutely the best person to turn to when estimating a change for *that* system. However, if you're building a system in another business domain, you're probably better off consulting someone who's worked on one project per year for the past ten years. Not all years of experience are equally applicable—some people have the same year of experience ten times.

π or e?

Earlier I talked about having the estimate conversation with an engineer. There's another set of steps you have to take as well to get to a more accurate number. When you ask engineers how long a given task will take, they naturally think of the part that they are most aligned with: the coding part. I've had loads of conversations that started with an estimate of X, and then when I asked, "Does that include unit testing? Integration? System testing? Documentation?", the final number usually wound up being a number closer to 3X, often a little more. They're not giving me the lowball number because they're evil or stupid. It's because that's the part for which they feel they are solely responsible. It's where they have to take the lead; the rest is not theirs. When I can't get the time to walk through the details, I like to use pi (3.14) as a multiplier for engineering estimates I get from junior and most midlevel engineers. For more senior folks, I use the natural logarithm e (2.71).

Accounting for Unknowns

Next you have to refine that number further. The number you get back will most likely include the engineer's internal multiplier for uncertainty. They might believe that the work can be done in two days, but since potential things could go wrong, they'll tell you four days. You have to be very clear that the estimate you're asking for is the estimate to do the work.

Manager: "Bob, what would it take to do this if absolutely

everything went perfectly? What's the shortest time it could take, and what circumstances would that require?"

Engineers are rule-driven people and they need to be accurate, so they'll tell you.

Bob: "Two days."

Doing the math that even a manager can do, you've now been told that there are two days of risk on this individual two-day task. That's bad. 100% is not good. You need to explore that.

Manager: "Okay, so what has to go wrong to make it take four days? Tell me about that."

And they will. You'll learn all about their concerns and dependencies and things that could go bad, and you might want to cringe, but pay attention and take notes, because this stuff is golden. Move on to the next task and do the same thing. Often the same risks, worries, and dependencies will affect the next set of estimates. This is key as we'll see in a minute. By the third task, most engineers will be giving you two estimates—the raw value and the if-the-creek-rises estimate. Remember, they're good at pattern recognition, and they want to get back to their real work and stop talking to you.

When you're finished asking about the tasks, you can figure out the best-case and worst-case scenarios for your project...with the proviso that you're estimating a biased problem statement filled with the assumptions that I talked about earlier.

Handling Risk and Uncertainty Within the Project Schedule

Imagine it's first thing Monday morning. Good engineers will finish their tasks on time. It's a source of pride. In the example above, if all goes well the task will take two days. If I put it on the project schedule for two days, it will be done by the time the engineer goes home on Tuesday. Makes, sense no? Monday, Tuesday...two days later. If I expand that task to cover the two days of risk, then it's due on Thursday. When will I get that task turned in? Thursday. Doesn't matter if none of the bad things happened. If it's due on Thursday, I'll get it Thursday, but I lost two days. The engineer isn't lazy or irresponsible; the task was due Thursday and he's right on time. He probably helped someone else out with something they were stuck on. He might have done some research on other things down the road, but either way, I'm out those two days.

Instead of adjusting the duration of each individual task to account for the risk and uncertainty, I prefer to handle that via a separate risk task. You don't use the risk task unless you actually have to. You should consider including a risk task for each logically related portion of the system, for each set of tasks that might share a set of risks. Once the risk has passed, you get that time back. The engineers know that they are covered should things outside of their control happen to the project. There's time in the risk task. At the same time, we're all working to a set of estimates that we developed that we believed were correct and have every reason to be able to accomplish on time.

Another advantage of this approach is that it puts risk management back where it belongs, which is right smack in the middle of everything. You're staring at a block of time in your schedule that you can get back if you actively manage the risk away. You can't ignore it.

How big a block should you include for each risk? I like to use an approach similar to an operations research method. I look at the probability of the risks manifesting and the durations and then apply some Kentucky windage to it. For example, if I have a 50% chance that a risk might happen and it would cost me three days if it did, I might add a two-day risk task to the schedule. If I also had two other risks that had a 75% chance of resulting in three days, and a 25% risk that could require 1 day, I might bump the risk task up to three days total. If they all happen, well, I'm in trouble, but usually they won't all happen.

Clients look at your risk tasks with a lot of confusion, worry, and angst. It's very important to talk through the whole concept of the risk task. It's not padding; we're working to the actual task estimates, but if these bad things happen, we've got it covered. If they don't, we can work to include additional features that weren't on the original release plan. Those risks, if they don't happen, are the gravy to the project. They are treats to the users.

What you must never do, though, is to add risk days over and above your funding. By that I mean if you have 30 days of funding, you must not have a schedule with 30 days of scheduled tasks and then an additional six days of risk tasks. The risks are real days. Your scheduled tasks plus your risk tasks must not exceed your total funding (in this case 30 days). Risks don't come free. There must be funding to pay for them.

A Few More Thoughts on Estimating

I described earlier how most projects get estimated, by one or more gray beards who looked through the available documentation, made their analogies, and with or without some model came up with a number. It's a bad idea to share that with your team when asking them to come up with more detailed estimates. Don't do it; it really irks them.

To the engineer responsible for developing the system at hand, the opinion of an old guy in some other part of the company who doesn't have to actually build the system is irrelevant and can only cause irritation. If your engineer believes it can be done in less time, the gray beard isn't so smart now, is he? If your engineer believes it will take longer, that gray beard has no idea what he's talking about. It doesn't bound, guide, or have any other positive input to the engineer's estimating process. If they're so smart, then they can get down from their Ivory Tower and code it themselves...if they still remember how.

It's Easy to Forget Coordination

If you have five or more developers, one of the developers will spend almost all of their time coordinating things, so you can't count them in the developer pool. You also should not assign them anything on the critical path because things will go badly if you do. They might be flattered by being the glue that holds the rest of the project together, but at the same time, they love to code and want to contribute. They will be frustrated by the exclusion, but you can assuage some of this by reinforcing how important their role is and how it has to go to them because they are the only one who can see the big picture and make it all come together.

If you feel you must assign the coordinating developer a task, make it some trivial, fussy task that has no major tasks dependent on it— nothing on the critical path. Most systems have features that the users love but are actually technically trivial to accomplish. The advantage of that is that you get both the users and the team appreciating the coordinator's efforts, which partly compensates him or her for not being a full-time coder.

Getting a Straight Answer from an Engineer

See if you've had this conversation before:

> Manager: "How long would it take to build this system?"
>
> Engineer: "I have no idea! We don't even know what it's supposed to do."
>
> Manager: "It's supposed to be a publishing system for stats data."
>
> Engineer: "Wow, that's clear as mud. What sorts of statistics? Can people search on it? Do we have to generate the stats? How is it supposed to be published?"
>
> Manager: "Look, here's what the contract says. They need a system to publish their stats, and it has to be user friendly and run in their current environment."
>
> Engineer: "Oh well, that answers all my questions. Um, two years."
>
> Manager: "Two years! Are you kidding me? We only have six months!"
>
> Engineer: "Oh great, well six months then."
>
> Manager: "Well, which is it: two years or six months?"
>
> Engineer: "You just told me I only had six months!"
>
> Manager: "Yeah, but what's it actually gonna take to build it?"
>
> Engineer: "There's no way I can answer this question. You have to tell me more."

The good news is that everyone is unhappy after this chat. Managers feel abandoned by their team members, who won't give them the information they need, and engineers feel trapped and set up for failure for not having the information they need. Hurray! There's a better way to have this conversation.

Management, if they are E-types, are generally pretty comfortable with uncertainty. They are willing to make decisions based on sketchy data, sometimes on nothing more than their gut instinct. They want to make the decision and move on. I-type engineers are not happy with this at all. They want data, lots of data, authoritative data, solid data. They want the data to lead them to the one and only one optimal decision. It is the only comfortable basis for a decision.

The managers were quite content with the vague request from the customer. They went ahead and said, "Sure, we can do that!" even without knowing exactly what *that* was. In a parallel universe, if someone was silly enough to let the engineers have the lead on a conversation with a client at this juncture, an engineer would have stayed in the room with the client for as long as it took, asking the client a thousand questions—all very reasonable but all very discomfiting to the client. That team would likely not have been awarded the contract. The "Can Do" team got the contract.

So how do you get an answer from an engineer when you don't have solid data for engineer-class decision making? You have to bound the problem.

> Manager: "Bob, I have a new contract for a statistics publishing system, and I need your help figuring out what we could do in six months."

> Engineer: "I have no idea! We don't even know what it's supposed to do."

> (notice: so far the engineer is defensive and feeling trapped)

> Manager: "Well, it's an insurance association, and I suppose they have all sorts of data on theft and casualty information, car repair costs…I don't know what else. What else might an insurance company care about?"

> Engineer: "Breakdowns by age and gender, average total claim, frequency of soft tissue injuries…"

> Manager: "Yeah, that sounds right. If we assume they have this data in something searchable, how long would it take to publish it?"

> Engineer: "I have no idea. What do they mean by publishing? Big thick books? PDF reports? An interactive web page?"

> Manager: "I'm not sure yet. Which would take the longest?"

Engineer: "Well, that depends. Publishing a real book is a pain, and we haven't done that before. That could take a while. We'd have to either find some package that does it or build it ourselves. Either way I don't think it's happening in six months."

Manager: "Wow, okay, let's assume we're not doing books on paper. What about the interactive website. How long could something like that take?"

Engineer: "Er, that depends: how many data sets, how many views, what sorts of breakdowns they want. When you say searchable, does that mean we have to search the raw data and generate the statistics or are they already calculated?"

Manager: "Let's assume that they've already calculated them and you just have to pull them up. And let's assume we have four different kinds of data, health, automobile, casualty and theft, and fire."

Engineer: "Well, that depends…"

I think you see how this conversation is going. You need to help two people: yourself and the engineer. You need to help yourself to understand the technical limitations and boundaries of what you've gotten yourself into. Did you know that book publishing was that big a pain in the neck? Did you know that the interactive website was easier? Did you know that the classes of statistics were a key input to your estimate variability? You need to learn all the gotchas that the marketing team never bothered to ask.

You need to help the engineer think through and identify those variables. In the process you're taking some of the stress out of the conversation by allowing the engineer to set out his or her concerns and then let you make some gut instinct assumptions on each of these concerns. The manager might have had no idea what classes of data are in the client's data warehouse. For that matter, you might not even know if there *is* a data warehouse. You just said, "Let's assume it's in something searchable." Boom! A whole area of concern removed from the engineer's estimate and another layer of angst removed from his or her thinking.

The first sentence is actually key. The manager changed the very nature of the question by flipping it around. What could we do in six months? We've also not asked for the system in six months, whatever the system might mean. Instead we've asked for how much could fit in a six-month basket. We've engaged the engineer in deciding what is possible. Now right off the bat, the engineer will have no idea of an answer, because we still haven't explored the boundaries of the problem, but at least you've asked for help solving a puzzle and not set up the engineer to be wrong and fail on the project.

The manager also engaged the engineer in determining what possible sets of data might be interesting to an insurance association. Never assume that engineers have no knowledge outside of code generating. Engineers read and explore things all the time. You would be amazed at the depth and breadth of knowledge they possess. Sure, sometimes it's Scottish heraldry or Klingon idioms, but there are also a slew of other things that they are interested in. And if they're interested in it, they've researched it, because again…they hate to be wrong. One day I was surprised to learn that a female engineer I worked with had an encyclopedic knowledge of professional boxing. You could have knocked me over with a feather. This wasn't useful to our project, but it shows you how broad their interests can be outside of engineering.

While no one much likes being wrong, engineers hate it. If you pride yourself on being master of your universe and able to manipulate things in the world, being wrong is really bad. Managers are wrong all the time—about as often as the weatherman—but they adjust and move on. No one seems much to notice the number of times that a manager is wrong. They notice the overall cost and schedule, but they don't harp on each individual error. When an engineer is wrong, it gets logged in the defect tracking system. It's reported out; everyone gets to see it. It's embarrassing.

So to avoid being wrong, the engineer will resist answering any question unless it can be fully specified. Lacking that, they will caveat it into oblivion.

> Manager: "How long will it take to build this statistics publishing system?"

> Engineer: "Well, as long as they have a SQL-based query tool and all the statistics are precalculated and there's only a single class of data to be displayed and we don't have to show the

raw data and we can reuse the query tool we built for the Acme project and the logic layer from the Sinclair job and they operate in a .Net world, then it should be doable in six months. Provided we get Bob to handle any customizations in the query tool and Andy for the Sinclair logic layer."

Responses like this irritate the heck out of managers. They have no idea if even one of these caveats will hold. The number the engineer has just given them is absolutely worthless. Why can't they get a straight answer from these damned engineers?

The answer is they *did* get a straight answer from the engineer. The engineer provided an answer that was accurate. The problem is the manager asked the wrong question. To the engineer, you've just asked, "How long will it take to build a house." Well, what house? A big house, small house, manor house, outhouse? Do I have to buy the land, build a road to it, include solar power? Is there water? Sewer? Electricity? Tudor, colonial, Italian villa? And it matters because the engineer wants to be right, and management is going to take their estimate, assign a cost, and hold the engineer to that cost. They always struggle to be as accurate as they can be. You've handed them a question with no answer. So they give you an answer with a fully stated question. "Well, if it's a three bedroom, two bath, center hall colonial with a total usable space of 2,000 square feet, vinyl siding in an existing neighborhood with average materials, it will cost $250,000, give or take $20,000."

Being wrong or imprecise is bad. Engineers hate to be corrected or even have their statements refined. It's why their answers are either one word or fully qualified. If I said, "The vowels are A, E, I, O, U, and Y," I promise you an engineer would correct me with "and *sometimes* Y." That is the more precise and correct answer. Are they trying to be a jerk? No, it's a gift really. Seriously though, no, they are trying to be more correct. They are trying to ensure that their understanding of the universe is full and complete. They were taught that "Y" is only sometimes a vowel. Now you go up there and say it *is* a vowel. That's not correct; you've said something contrary to their world model. Y is only *sometimes* a vowel. The next sentence coming out of your mouth might be incorrect if it is premised on the notion that Y is always a vowel, so just hold on there pardner: reset! Remember whom they speak with most of the time—the computer.

Remember how moronic a computer is. Remember how incapable they think someone is who can't control a computer (i.e., you), so yes, they need to correct you because, well, you're not all that much more intelligent than the computer.

As an engineer, it's embarrassing to be corrected.

> Engineer A: "You'll never have a black swan."

> Engineer B: "I saw one at a zoo in the rare and exotic animals section."

> Engineer A: (sighs)…"Okay, you hardly ever get a black swan."

> Engineer A: "We've always done it this way."

> Engineer B: "Well, except for that project five years ago when we didn't."

> Engineer A: "Okay, so every time but once we did it this way."

Recall that engineers need to be very precise to do their job, and they have to constantly think about the exceptions. It's what makes them good at their job. Yes, in this situation it's really annoying, but you can't ask them to switch it on and off like a light any more than you could ask them to change their hair color every day of the week. In each example above, although Engineer B's exception is moot and perhaps absurd to the conversation, Engineer A lost face/coup. He felt a pang of momentary uncertainty because he was not fully in command of his facts. It will pass quickly, but he will feel unhappy.

If at all possible, help bound the problem with a similar project.

Manager: "I think this is like the Acme project but with about half as much data, and we'll need to display it in about twice as many ways."

Boom! At least the vast universe has been reduced to our own solar system.

Schedule

The primary use of a project plan is to give the client something to bludgeon you with at every meeting. That's not its purpose, but that's its most common use. I wish that were only a joke. The project plan can consist of a huge number of documents depending on the type of client and project. The Project Management Institute's body of knowledge includes the following:

- Scope Management Plan
- Schedule Management Plan
- Cost Management Plan
- Quality Management Plan
- Staffing Management Plan
- Communication Management Plan
- Risk Management Plan
- Procurement Management Plan

And these are just the sections of a Project Management Plan. There are other documents to boot!

No matter what you call them, there are four things you have to figure out on a software development project.

- What are you going to create?
- How are you going to do it?
- What do you need to do it?
- In what order must things be done?

One thing I agree with the textbooks on is that the project schedule is built around the things you are going to create, not the steps to make it. The top level is the list of products that will come out of the project. In some cases it might be a category of product, but it is never an action. If you list the actions, you lose visibility into the products you're required to produce. They become incidental. For example, I've seen scads of projects following a typical waterfall development cycle that look like this at the top level:

Wrong:

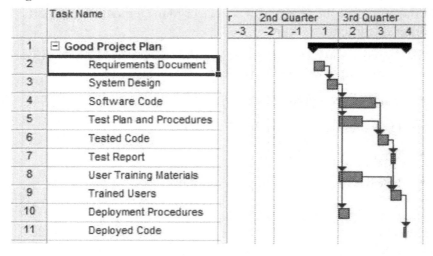

	ⓘ	Task Name	r	3rd Quarter			4th
			1	2	3	4	5
1		⊟ Bad Project Plan					
2		Analyze Requirements					
3		Design System					
4		Code System					
5		Test System					
6		Train Users					
7		Deploy System					

Figure 1. Incorrect Task Naming

What are we creating? A system, sure, but what are the things that need to be developed, tracked, delivered? They're all lost in the doing that is being listed.

Right:

	Task Name	r	2nd Quarter			3rd Quarter		
		-3	-2	-1	1	2	3	4
1	⊟ Good Project Plan							
2	Requirements Document							
3	System Design							
4	Software Code							
5	Test Plan and Procedures							
6	Tested Code							
7	Test Report							
8	User Training Materials							
9	Trained Users							
10	Deployment Procedures							
11	Deployed Code							

Figure 2. Correct Task Names

Notice how soft and fuzzy the first list is compared with the second. Notice how the first is friendly and nonthreatening and the second is hard and cold. Hard and cold is what you want. I know we've been talking about the human side of things, but the schedule is one area in which you need to be rigid and unyielding. It's all outcomes, not process. The project plan has to be an outcome-oriented document. It doesn't matter how we get there as long as we get there. This is what we've all agreed the project should produce, and you must focus on that.

Where do you get the list of products to be produced? This is usually pretty obvious. If you're a contractor, it was listed in your statement of work. If you're a federal contractor, it's the Contract Data Requirements List, or CDRL. If you're an in-house development team, it's the required documentation for whatever your internal process is. The items at this top level become the most visible parts of the project. They are what will be discussed at every review meeting. Not the analysis meetings, not the design sessions, but the products. They set the expectations of the project. If it's not on the schedule, you're not producing it. In fact you need to set that expectation right from the start. You need to say out loud at the very first meeting to review the schedule, "This is the list of the products we'll be creating through this project. They came from the contract Statement of Work (SOW). These are *all* the products we'll be creating." You will have to repeat this on occasion, but you must establish a common understanding of what is being produced right away.

Be very careful about listing products that are *not* in any of these documents. If you do, you are adding to your workload. If you know that the client needs something that was not included in the contract and you did *not* add it to your bid, then you've already made one mistake. Do not make a second by adding it to your schedule without renegotiating the effort.

I've had customers who honestly did not know what documentation they needed from all of the various parts of their organization. Some didn't know what specifically would be needed for this system, and some couldn't have pulled together the list if they had to. They just didn't know what their own process needs were. They tried to hide this by including in their list of required documentation things like "Standard Operations Documentation" or "Deployment Materials." That's fine. They just don't know, so you have two choices: (1) get them to enumerate the documentation they need, which they would have done in the first place if they could have; or (2) list what you think is needed and negotiate from there. Option 1 will annoy them, so I recommend option 2, which reinforces the need to say out loud,

"This is the list of products we'll be producing. It's what you bid in the contract or put in your budget." If they say what they meant by "deployment materials" is a deployment review packet, then you have to push back if what you bid was a set of deployment procedures.

If you agree to produce additional products within the same cost envelope and/or schedule, think about what you are telling your customer:

1. You pad your estimates.
2. You pad your schedules.
3. You are ripping them off or "sandbagging."
4. They should not believe any number you give them.
5. This project could have been done a lot faster and cheaper.
6. They should keep asking for more products since you've padded everything so much.

First of all, you shouldn't be padding your schedules and budgets beyond a reasonable management reserve to account for risk in the items you bid. You should never be ripping off your customer, and you must always tell your customer the truth. So, on the assumption that you are following those rules, any new product that you did not include in your bid (or budget if this is an in-house project) and is not on the schedule *must* result in a renegotiation or you will fail.

Every product block in the schedule needs to have a subtask for delivery of the product, that is, Delivery –ProductName-. This is then followed by a subtask for formal acceptance, or Client Signoff/Acceptance of –ProductName-. You must establish the expectation that once you produce the product, they have to accept it and own it. It must be *their* product, not yours. You're on the hook to create it, but it's for them, the customer. It's not something you did because you thought it was fun to do. It has a purpose, and the purpose is the client's, not your own. Making your client formally accept the document sets the mindset that they must pay attention to each product, that each has value and that the value is to them, not to you creating some system for them.

If all they care about is the final system and don't give a hoot about the intermediate steps (products) it takes to get to a quality end product, you're doomed. How could you be expected to build a perfect home for someone who won't review the blueprints with you, who won't look at the color swatches for the drapes or the samples for the tiles? When building a house, most people can relate to the things that go into it, but for software everything seems arcane, strange, and uncomfortable.

It's your job to make it clear to the client why each of the intermediate steps adds value and how each step contributes to that value. Otherwise you'll have months of time when your client is just nervous and anxious and has no clue about what you're doing to justify your invoice. You won't necessarily see it, but that anxiety just grows and grows. You may assume that things are great because you're building all the intermediate products, you're on time, and the documents are clear and complete. You think life is great, but it isn't. The customer doesn't understand what these documents mean to the end product. To them, they're as meaningful as the rosin mix in the plumber's solder. Why should they care?

You get closer to the end and finally show up with some software that runs on their machine, and they now get to look at it. It's not what they wanted, they knew it wouldn't be, and to heck with you and all techies! Walking your users through the steps in the schedule and explaining how each step supports the end product are the keys to making sure they know what they are reviewing and how to look at it. An architect will walk a client through the blueprints, explaining the symbols and what they mean. With this knowledge, homeowners can then look and see if they think there's not enough room between the sink and the door or the light is too close to the mirror. They will know that if they don't decide this now, moving a doorway once the house is built is a major problem. If the architect does not explain all this, clients are just looking at some squiggles on a large piece of paper.

You have to decide what needs to be done to create each of the products. In many cases you're required to deliver a draft document, get customer input, and then create a final document. To create each, you need to take some steps. You need time for research, maybe meet with stakeholders, write, do an internal review, edit, seek internal approvals. So what goes into the schedule? Some simple guidelines:

1. Anything that the customer has to do *must* be in the schedule.

2. Anything that requires a decision point *must* be in the schedule.

3. Anything that takes more than a day *must* be in the schedule.

4. Anything that has a dependency *must* be in the schedule.

5. Anything that should be reported *must* be in the schedule.

6. Any task that takes five days or more should be decomposed into smaller tasks. You will be preparing weekly reports. You can't have a task that takes a whole week, because by the time you know it's going badly you won't have time to respond.

The key is to remember that you need to track the status of, follow-through on, and report exceptions to everything on the schedule. If you include tasks that take 20 minutes on your schedule, you can count on spending the rest of your natural days buried in MS Project and writing exception reports.

Dependencies

What about interdependent tasks? For example, in a classic waterfall, you can't start design until you have agreement on the requirements. So there is a true real-world dependence on approval of requirements to the design going forward, and that should be included in the schedule.

I was called in once to help a company figure out why they were 18 months late with the latest version of their product and still didn't have a clue when they would be able to deliver. There were a slew of reasons, but one of the obvious problems was that they had no idea how the schedule worked. I glanced at their Gantt chart and noticed there were no dependencies. *Nothing* had a line connecting it to another task. From a Gantt rules of order perspective, what this means is that nothing had any link to anything else. *Every* task was completely independent of any other task. They worked hard in MS Project to avoid adding dependencies, which is actually difficult to do. I knew the answer, but rather than call them stupid I asked if it was true that there weren't any dependencies in the schedule. They replied, "Oh we don't like dependencies."

No one likes dependencies, but here in the real world they exist. You have to have the code finished before you can deliver it. You have to have a methodology for testing it before you can test it. You have to know the expected results before you can design that testing methodology. You have to have a purpose for the system before you can list the expected results. You have to have some sort of a technical architecture before you can decide how to do a more detailed design. You have to know what sort of building to make and the allowed materials before you can engineer the structure. Stone? Wood? Steel? House? Warehouse? Gymnasium?

Dependencies are real. Some are temporal, some are related to resources, some are political, some are process driven, some are organizationally driven. There are many.

A common mistake is to list as dependencies a sequence of tasks assigned to an individual. Task A is dependent on task B, but only because Bobby is assigned to them both and you want task A to happen first. This is not the type of dependency you should enter into the project plan. That's the purpose of MS Project—it will figure that out. You need to include the dependencies that are not person-related. Must a committee change some policy before you can deploy? Must you get approval from security before you can deploy? MS Project knows nothing of the business interdependencies. You must identify them for the tool.

Doing this also helps your client understand the actual workings of the project. Although you don't want someone micromanaging you, you do need someone to understand why you're agitated about something that they don't know is in the critical path. How can it be that receiving a new switch (when we have loads of old ones) is worrying you so much? Why is a delay on someone else's project making you such a pain in the neck to deal with?

You probably have loads of external dependencies you're not even aware of yet. The operations team may have its own set of projects and can't get to your tasks until they finish things in *their* project schedule. For example, if they are upgrading to new servers and you need a development environment, you now have a choice: (1) stay with the old servers and pay the price down the road to migrate, or (2) wait until one of the new servers is ready for you. How do you find out which option is better?

The answer goes back to the fact that you have to spend a lot of time talking to a lot of people if you want to have a successful project. You need to be talking regularly with the operations folks. Is the training center going to be recarpeted and repainted, maybe rewired, or have new projectors installed? Maybe the new work stations aren't going to arrive until later in the year. Your training tasks now have a dependency. The security team might be scheduled to attend the Black Hat conference for a week, the same week you need them to review and approve something. Now you have a dependency (arguably a constraint but let's let that go for now). The users themselves might have a business process that takes up every moment of their day the month of October. Until that finishes, you can't get done all the things that you need them to do for you. Again, how do you find this out? You have to talk to people.

Many process-oriented people, especially those in a project management office (PMO), will push hard to have integrated program management tools so that everyone's projects throughout the organization are in a single tool and these schedules are integrated; changes in one schedule ripple through everyone else's. They view this as a good thing. I disagree.

It's a great thing if your job is to spend every waking moment reviewing and updating your schedule. However, as I've already beaten to death, that is 10% or less of your job. The last thing you want to do is open up your schedule, see that you're now going to be two months late, and then start playing "Where's Waldo?" to figure out which task has changed.

The tools can squelch interproject communication. Every project manager should know what other projects are relying on them for, and they should be talking to the other projects whenever they hit a snag that could cause a delay. People have to be in the loop, and they should be the ones to first figure out the impact and come up with ways to mitigate the impact before it ever gets raised up the management chain. What often happens is the PMO is the first to discover it, because the PMO has some weenie whose full-time job is to sift through MS Project Server and tell you that you have a problem. Now the PMO is not an asset but a source of irritation. The connections between the projects have to be owned by the individual project managers, not a PMO. They have to believe they are beholden to these other projects. They have to feel a responsibility to the other projects, and the PMO removes this in almost every case.

So instead, I recommend including a placeholder task for outside dependencies, for example, "New Servers in Place," and set it as a milestone with the date you get from the operations project manager for the server replacement. Don't try to link to their project plan file directly. It will only frustrate you and tie them down in terms of file location, name, task name, and the like. Keep it simple and rely on human interaction to keep the projects connected.

The Project Schedule World Tour

Great! You now have a schedule. It's resourced, has dependencies, durations—all the required stuff. If you did it right, you got input from the engineers on the team, but it's still a project schedule. It's a management document that none of your engineers cares a wit about. The thing you're holding in your hand is something you own. No one else is really very much interested. They gave you some estimates, not a schedule, so that's *your* schedule, not theirs. They haven't committed to it. What often happens in the next step is that you take the finished schedule to review with the engineers. You've already been talking about it for a while now, so this meeting will go well!

No, it won't.

Engineers look at schedules because you make them look at schedules. You asked them for estimates and risks and they gave them to you. Requested data were passed on to you—task done. However, it's not often clear to them what the result of those estimates will be. They might look at the overall schedule—the start and end dates—but they don't generally get a lot out of them. You need to introduce the schedule properly. Walk it through. If you have some very junior folks on the team, you may actually have to explain how to read the schedule. What does a milestone look like and what does it mean? What is slack and how is it represented on the chart?

Since you've already gotten the estimates for the individual pieces of the project, the goal now is to make sure everyone agrees on what the order of battle is and the dependencies. You'd be surprised to find out how often seeing the schedule's Gantt chart helps engineers see and identify dependencies they were previously unaware of. You'll get the "Oh, but we need this to be done before that can happen." Not sure what sort of cognitive processes suddenly fire when looking at the picture versus all those conversations you've had before, but it does work. Odds are you'll have to go away, redo the schedule, and come back to work out some resource contention problem. But don't worry, because you get to play this town again later on, after you've toured the schedule with other stakeholders. And who are those other stakeholders?

The End Users

The end users have a major role to play in a project. Even in the classic waterfall, the users have to be around, at a minimum, for the requirements analysis phase to review screens and do user acceptance testing. You need to make sure they will commit to being there when they say they will. If they are going to be reviewing a document and you have a block of time allocated for that, you need them to say out loud—or even better, in writing—that they will complete their review and provide input by the end of that block of time. It has been my experience that the number one reason for projects sliding to the right (i.e., falling behind, moving to the right on the Gantt chart schedule) is a delay in response from the end users. While you are busy writing things up and talking to other users, the other users are busy with their normal day jobs. They aren't thinking about how their day job has a major crunch coming up right when you want them to drop everything and read your silly document. It falls off their own work schedule.

You need to be explicit about what will be happening on your project. "I will be sending you an electronic version of the requirements document at the end of the day on Tuesday the 15th of November. You have three work days to review and comment on it. That means I need your written comments back by close of business that Friday, the 18th of November. It can't be any later or the project will slide day-for-day until we do get it."

They will almost always immediately say "That's not a problem." I strongly urge you to follow up. "Great, now you're sure that nothing else you're working on will be interfering with this. I genuinely think you'll need all three days to read through, digest, and comment on the document. It's going to be about 100 pages of very boring stuff that needs a lot of detailed review." This is when you'll get a panic reaction and some choking and coughing. "Oh well, my word, I had no idea it would be so large! Three days, wow really? But sure, that should work."

Notice that they often get stuck on the volume of work; they haven't said out loud that they can drop everything else for three days and review this document. You have to follow up again. "Great, now you're sure you will be able to dedicate those specific three days to this review? There's nothing else you're scheduled for on those days?" And here it comes: "Oh, yeah, well, that's the second Friday of the month, and we have to get our revenue estimates in before 3 p.m." Or something similar.

It's amazing how often the business users are unable to link up their normal work activities with what you're showing them on a project Gantt chart. You genuinely have to be this much of a nuisance to help them help you spot the problems. In fairness to your end users, you are showing them your project schedule. You're not showing them your project schedule with their activities already superimposed over the top of it. Now you've discovered a new constraint on your project, so you get to go back to your desk and redo the schedule to accommodate this latest information and then reschedule the review.

One other point about an end user review: do not assume that they work with or know how to read a Gantt chart. Many of them do not, because it's not something their normal work requires. It's worth discovering whether they understand it before you charge ahead.

Training Department

If you're bringing in a new system and you've got a dedicated training department, you will need to make sure that they can support you when you need supporting. If you are not a veteran of that particular department, there's a good chance you don't know what their process and dependencies might be. You should sit down with them and go over your schedule. The best way to get a gasp out of them and watch their skin go pale is to start with your estimate of training time: "We do a lot of analysis and design, then we code and test, and here is when we start the user training. I've got 100 users to train, so I estimated four classes, two hours each, with 25 in each class. We can get it done in one day. I have training scheduled for December 25th. Will that work for you?"

All kidding aside, depending on the organization, there might be all sorts of internal reviews, management approvals, and documentation to complete before you're allowed to train the users. Some of those approvals are not the responsibility of the training department, but training still can't begin unless they get them. They might only have the ability to schedule 15 people in front of computers at the same time. They might be relocating the learning center right around the time you want to have classes. They might have other classes scheduled at that time. They might think you're crazy to believe that average users can be trained in two hours. They might know that you only get a 75% attendance even if it's scheduled way in advance, so you'll need to plan for make-up classes. The point is that there is a lot of good input and insight you can get from the training folks. They will likely want to start engaging with the project around the time the requirements are finalizing. They will need to be around during testing to familiarize themselves with the system. They may need multiple sets of test data so they can have the trainees do some exercises. They will need some way to reset things after each class (a new development task). In some organizations, the trainers are the ones who create the help screens and user manuals if such exist.

You need to learn what those who teach need from you and vice versa. It is very common to dismiss the complications of training. "Since it's a nontechnical task, it must be easy" is a common trap to fall into. If the users are well trained, it will go a long way toward their acceptance of the system in those early days. If they don't have a clue how to use the system, they will complain to your help desk and their manager, and that will all flow your way. So making sure that the trainers have everything they need and the time to do their jobs well is definitely in your project's best interest.

I've never had a schedule review with a training department that didn't require me to return to my desk and do some major work on the schedule. You will get to come back to them to review those changes.

Operations

Your review with the operations team is extremely important, downright critical to getting the project started. If you don't have a development environment on the first week of the project, you are often already late. Usually operations was not included in the discussions that led to your project getting approved and rolling. They were busy doing their normal work of heroically keeping the systems up and running despite faulty hardware and vendor software, intermittent internet service, hacker attacks, less-than-brilliant users, and other travails. When you called and asked for a meeting to review your project's schedule, they were less than thrilled. Odds are some fire is burning right before your meeting time, so they will be distracted by that while you're trying to explain why you need something special from them and you need it in a very short amount of time. Do not expect a warm reception, but do understand that it's not personal. They hate all new projects.

My recommendation is that you do not start with the schedule. Instead start with a description of the nature of the system. "This system will be about as big as Legacy System X. It will have almost the same size database; in fact it will be reading Legacy System X data. It will have a smaller user base; only the reporting analysts will be using it, but it will be very CPU intensive. It will have to be up during normal business hours only, no 24/7/365 requirements..." Next, talk about when you will need a development environment, a test environment, and when you plan to go operational. Don't forget that some organizations also have a training environment, so there are a slew of potential servers and databases that operations will need to set up for you.

You probably won't get all the way through this before they start asking you what the environment needs to be. If you know, be prepared and tell them, "This will be our normal LAMP stack (Linux, Apache, MySQL, PHP) with the connection to the Legacy System X Oracle database." If you don't know what the environment will be, be prepared for some major frownie faces and possibly some comments about your intelligence and the value of your project. You're saying you need an environment set up pronto but can't tell them what that environment is. You'd frown too.

This schedule review is a great time to try to include the operations team in the effort. Once they're past the initial shock of the project, indicate when you'll be ready with draft documents. They will surely want to see the technical architecture as soon as you have it ready. They will want to know when you're closer on the database size estimates. You might want to invite them to other reviews just so they know more about the system they will soon be responsible for supporting. Please invite them. They might not be able to attend, but invite them anyway. The invitation makes them feel included and part of the team.

From the operations team you will also find out about all sorts of things that the enterprise does that might have an impact on your schedule. The accounting department might be closing out the books, which will require operations to set up some temporary set of archive servers. The accounting data you think you'll be using for testing or for actual operations might not be synced at that time. If your client is a separate business unit, they probably have no idea this goes on and never mentioned it to you. It could be that August is when all the licenses for some infrastructure product the company uses have to be refreshed, and it takes a full week to ensure that everyone is using the same version. In the past they've seen individuals who, for whatever reason, would not let the license update go through and then wound up unable to do certain things. This kind of complication can affect your users' success in working with your system.

Operations can tell you if you will have to make a special request for the continuity of operations environment. Many companies are now setting up a way to keep things running in case the main computing environment is unavailable because of some sort of disaster or other reason. The problem is that most companies are still new at planning for this, and the process and procedures and policies are still little understood. It generally has fallen to operations to figure it out; therefore, they are the font of this knowledge. Pick their brains for potential problems you could face. They know more about the entire enterprise's business cycles than you might imagine.

Once again you will need to go back to your desk and redo the schedule and then come back to them with the changes.

Testing Department

Testers, like any group whose tasks fall on the right side of a schedule (trainers, technical writers, security reviewers), are used to having their time cut and getting things late. They hate it, but it comes with the territory. You call to set up a meeting to review your schedule, and they probably first smirk and think to themselves, "Oh look, how cute! Here you are with your brand new schedule. It's so pretty! And it has all sorts of time for us to do our job. How thoughtful! Wonder when we'll actually get it? Bet you wind up giving it to us with two days left, to do our ten-day job. Oh, goody. Looks like the end of June is going to be a nightmare."

As I mentioned earlier, a good testing department is a wonderful thing. You would be wise to give them the time they need to help you deliver a high-quality product. If you skimp on testing, you are only adding risk to your project. You don't want your users to be the first ones to find an error.

The testing crew has a methodology of their own to figure out how to break your system. That's their job. They will need to be involved, starting during the requirements phase. The requirements will tell the testers what the system is supposed to do. They will have to make sure it does that. They also need to figure out what might go horribly wrong, which takes time. You need to include them on the schedule for a requirements review. You need to include them on the schedule for a design review. You need to include them on the schedule for a user interface review. The more they know about the system, the more completely they will be able to break it. This is a good thing.

Your testers are also likely your first sources of information about the latest set of intrusion detection and security vulnerability tests that need to be performed. The first people to learn of this were on the project that was supposed to deploy on Tuesday, only to learn on Monday of a new requirement laid down by the Information Security folks. You can learn about it earlier by keeping in touch with the testers, since they are the ones who most often actually run the security testing software. It may require some sort of testing harness set-up that takes a while to complete, the time for which needs to get back into your schedule. If your system must be compliant with Section 508 or other accessibility standards, the testers are likely the best bet for figuring out how to ensure that and how long compliance testing will take.

Guess what? You're headed back to your desk to redo the schedule and re-review it with them again.

The Encore Tour

You've probably detected the pattern of review-redo-review again. You've probably also guessed that sometimes one organization's constraints or processes will change tasks and tasking for other parts of the team. For example, the developers probably didn't know they'd need to build some script to reset the training environment's database between each class. When you pull in all the added tasks and dependencies, the schedule can change quite a bit. Not only will it be a lot more complicated, with little dependency lines going this way and that, but it will be longer.

Your engineers, in particular, will be unhappy and unsettled. They were probably already thinking about the tasks they were going to do. Without even the first bit of a requirement, they were probably scheming about how to build this system. Now you've come back to them with a whole set of ancillary tasks that were not part of their mental model. They won't like that.

My suggestion is to begin by acknowledging the irritation of the change. Remember they are judgers, and you're messing with the plan. "Bob, I need your help. I was talking with the training department, and they pointed out that they need a dataset to run the users through some exercises during the training course. You gave me an estimate for that, and it's already in the schedule. What they reminded me, and it's my fault for not thinking of it sooner, is that they need to be able to reset that database before each class. Is that something that's hard to do?"

Notice I did not throw the training department under the bus for this request. Take the hit for not seeing a reasonable request. Keep their frustration pointed at you. By framing the request as an obvious omission on your part, you avoid the "Why don't they just do another set of exercises with the data as it is at the end of the class" discussion.

You also need to recheck that the changes in the schedule haven't moved some tasks smack into other constraints, like vacations or other business cycle activities.

With each of the groups, you should start with a thank you for their previous help and show them how their input was incorporated into the schedule. It's also a good idea to show them how other groups' activities have changed the schedule. The more everyone knows about the interactions and interdependencies of the teams, the better.

The Grand Finale

Once you have every group's input, you would be wise to set up a time to review the new master plan with your most senior project sponsors. Make sure you have your notes on why things that might not be obviously dependent on one another are set up that way in the schedule. This review is more than a courtesy call. You need the sponsors to bless the schedule. From here on, you will be referring to this as the "approved" schedule. You'll be reporting on progress against this approved version. If you need to apply pressure later, your only leverage point is that this is what *your* boss or his/her boss approved. Deviations from this are something that you're required to report, and they need to know why they are happening.

Schedules imply a budget, and you must make clear that this is the budget that goes with the schedule. The sponsors are approving that money and need to justify any changes to their own leash-holders. An increase in a budget necessarily results in a decrease of return on that investment, which is not something they will do casually.

It's also a final opportunity before the heat of battle begins to learn about any constraints or resource contentions that might exist. "Oh wait, you're planning the acceptance meeting for September 18th? That's during our senior management retreat. That won't work. No one will be here. We'll be on the links." I strongly encourage you not to let your jaw drop or your shoulders slump. Yes, after scads of meetings with every picayune group of people in the building and all the back and forth negotiations you now have to again change the schedule and add a week to the budget to allow these guys to play golf on the company nickel! It's good to be king. Since you are not king, smile and ask them when someone would be available for that approval meeting. Make sure to tell them that this delay will add X days to your schedule with the associated labor costs: "Okay, I'll move the date to the following Tuesday. This will add three days to our labor budget as I can't release the team for only three days." Sometimes they will look at you with shock and horror. Sometimes they won't even blink. Most often they will find some duke, countess, or earl who will be their proxy for this meeting and you get to keep your schedule as is. Phew!

The Structurally Impossible Schedule

Sometimes the dependencies, processes, approvals, business cycles, and fates make it impossible to craft a schedule that can be completed within the time allotted and/or the approved budget. What to do?

A consultant friend of mine had a contract for a job with a period of performance that ran from September through the end of January. Included in the work to be done was the end-of-fiscal-year initialization of the database for the following fiscal year. But the business does this in February. Tricky if your contract is already over.

I've had projects that had deliveries during peak business cycles when we could not interrupt the users for testing and training and acceptance and deployment. I've had schedules that ended during the week between Christmas and the New Year, when most companies don't have anyone around. I've had schedules that called for the needed new equipment to be set up in March, but that equipment had to come out of the next fiscal year's budget, the fiscal year beginning in May.

More common are schedules that are stretched out because of other priorities in the enterprise. The trainers can't support you that week because everyone is in the latest MS Office training as part of the big upgrade. The users are busy closing the books and can't go to training. There are probably a hundred other examples.

Sometimes you can come up with workarounds, such as using old servers until the beginning of the fiscal year. Sometimes you can get approval to move the data early. But sometimes you're just plain stuck.

The solution is sadly obvious. You need to set up a very structured meeting with the senior managers because they have a decision to make. You first have to explain the problem at the 100,000-foot level. "We can't complete this project on time and budget because we finish up right in the middle of the holiday season, and most of the people who need to be trained, and the trainers themselves, will be on vacation with their families." Be prepared to explain why you can't compress the schedule two weeks to allow the project to finish before the holidays. Give them options. "We can either extend the work until the second week of January at a cost of X dollars or drop the reporting module and finish the last week of November. I'll need to know which way to go by September 10th."

Do not leave it open ended. Management is filled with deciders and action-oriented people, and you don't want them dreaming up all sorts of approaches, the flaws in which you won't be able to explain to them in the time allotted for this meeting. Control your meeting as best you can by properly preparing and staying on message. If management comes up with a new idea in spite of your best efforts, and your gut is telling you it won't work but you're not sure why, then accept it only provisionally. "I hadn't thought of that. That should work, but let me check with my guys to see if I'm missing something. If not, we'll go with it. If there is a problem, what would our second option be?" You didn't tell the king no; you said okay, and now you're letting them show even more wisdom by offering up their next-most-favored option.

The IT Department

Let's talk about the IT department. In most every enterprise there's an organization responsible for keeping the lights on and the network humming. Sometimes this is in the same group as the development shop. Sometimes it reports to the CIO while the development team reports to the CTO. In larger organizations it might be further broken out to network, operations, and possibly further defined by platform: Unix/Linux versus Windows versus Mainframe. Regardless, no amount of emphasis is too much when it comes to stating how important it is that these people be on your side. They can and often are very willing to bring your project to a complete halt, and they have no qualms about doing this at the last minute.

IT owns the servers your system will run on. They own the people who monitor the health of those servers. They control who can access those servers. They own the network to get your users to the servers. They have a strong influence on what technologies can be used. They often run the security infrastructure. They are often the only ones allowed to manage a production machine. They control what sort of a development environment you get. They control what the test platform looks like. They control the process to move an application from development to testing to production.

So they have a lot of control. You don't.

First, let's take a step back and look at what their motivations are. The operations crew has one of the least enviable jobs in the IT sector. They have to make sure that flaky hardware, flaky software, flaky networks, and flaky power systems made by multiple vendors that may or may not be compatible never manifest their flakiness. Never. And they get no credit for doing this. They are punished when there are systems failures and ignored when they prevent disasters. Systems don't even have to fail for them to be punished. If things just get busy, they can expect a flogging for slow system performance. Since they most often report to the CIO, they are given as small a budget as possible. Most of it goes to huge enterprise licenses for technologies such as Oracle, Microsoft, or SAP. These big-ticket items are painful for nontechnical managers to accept, so having to budget for people to manage this technology is very hard for them to swallow. Corporate managers hear about how operations personnel are interchangeable (after all, we have multiple shifts) and could be outsourced, so why pay top dollar for them? Many organizations feel that the IT budget overall is a burden, a cost of doing business that is only reluctantly borne. No Unix system administrator ever signed the deal that brought in the bucks! No network guy ever came up with the new product that will make their bonuses come true. So IT is in a tough spot.

Management hates dealing with the operations team maybe even more than you do. Think about it, a typical conversation goes like this…and by the way, is only prompted by something that's not working (cuz why else would you talk to those jerks?):

> Manager: "What's going on down there? I can't get to my order system."

> Operations person on duty: "We know, we're checking into it."

> Manager: "Well, when will it be back up?"

> Operations person on duty: "We won't know that until we figure out what's wrong."

> Manager: "I have a meeting in 30 minutes, and I need those numbers."

> Operations person on duty: "Well, we know it's not a DNS issue because I can clearly see the server. It might be a problem with active directory. The AD guy is out with the flu, but Carlos is talking to him on the phone now. It might also be a file extent problem, but until we can get past the access side of things we won't know."

> Manager: <didn't understand a word of that> "Okay, but I need those numbers NOW!"

> Operations person on duty: "Well, maybe you shouldn't have waited until the last minute to get them."

> Manager: "Look, you're supposed to be a service organization, and this isn't acceptable!"

> Operations person on duty: "I don't know what to tell you. The system is down, and we're working on it. Do you want the ticket number?"

> Manager: "You're kidding me! Who's your manager??"

The conversation, already bad, will not get better. Odds are the IT supervisor was already leaning over someone's console trying to get the problem fixed and will now be dragged over to talk to an angry business manager. Swell.

The operations teams must maintain skills in operating every type of technology that runs in the company. They need to have people on shift with all these skills every day, and most nights. Users want every new thing under the sky because they saw it at their buddy's house or on TV or in a store. It's so much cooler than the company-issued device they have, and they want it now. Let's assume the best of all possible worlds: it is compatible with the existing systems, it will securely connect to the infrastructure, and it can't create any new security leaks. Operations now has to learn all there is to know about this device before they can support it. Same sort of situation crops up for every development project. There's a new library, a new add-on for the web server, a more efficient language, so the developers want to be freed from the constraints of the old stuff and use the new cool stuff. If they don't use the new cool stuff, their skills will be aging, their value decreasing.

Senior management will often insert new technology by fiat. So management and all the users see operations as an incompetent group of belligerent, overpaid dilettantes, and they sure as heck don't want to deal with them. They just want this new accounting system, and well, operations is going to have to get up off their whiny prima-donna butts and learn a whole new environment, and *no*, they can't have any more people to make it happen.

Operations can't do it all, so they say no. They are in the business of saying no. They have to. Very few of them enjoy the frequent floggings, and even if they did, they just can't keep everything and anything up and running in a reasonable fashion if they say yes every time.

Operations is the abused spouse of every enterprise: never appreciated for the days things go well, unrecognized for disasters averted, always dumped on, and always working hard to correct the problems created by others. So here you come along with your cute little project. From their perspective, this isn't a new tool to help the enterprise; it's just one more thing that can go wrong and get them their next beating. They have no idea what your system will do to them, but they sure as heck know it won't be anything good. There is nothing that adding another business application can do that would improve or simplify the operations team's life. That's just a fact.

Most operations teams have some sort of service level agreement, either explicitly or tacitly, with their enterprise management. The SLA is operation's contract with the company to keep things up and running at a stated percent of the time. Operations is rewarded or punished based on how well they hold to this agreement. Every time you come in with a new application, you jeopardize their SLA. So if any change can only cause you to miss out on a potential promotion, raise, or bonus, what would you do? You'd resist change…vehemently.

My point is not that the operations team is evil. My point is that we've incentivized them to make your life a living hell. So you're a threat to the operations team, and they control your ability to be successful. Great. What do you do about it?

One might think that the solution is to get the CIO, COO, president, or Thor to use a mighty hammer to mandate cooperation. That's a solution but not one with a high probability of success. You started as a threat, and now you've leapt up the hierarchy of bad to the source of a pre-beating for sins not even committed yet. They'll love that. This approach is the nuclear option and should be attempted only as a last resort.

The operations team is a major stakeholder and has to be included from the start. In an ideal world, they'd be included in the initial planning of the project. They should have been there before you were there. They should have been providing input to the concept of operations, the technical requirements, and more. But odds are, no one wanted to deal with "those" people, and *you* get to break it to them. So get on their side. Their goal is to minimize their pain and suffering (remember you can't be a benefit). Okay, if your application is replacing something that was nothing but trouble for them, you *might* be able to break even, but in their minds this system is bad news today, and you're just going to be the new form of bad news tomorrow.

How do you convince them that your system is not going to be trouble? You need to find out what's the scariest thing for them about your system. They might not even know it. Learn their schedule. Learn where their staff are weak and maybe your team could help. You could actually collaborate! Learn what they want for cutover. Ask if someone from their team can attend the reviews. Send them the monthly status reports (if not the weeklies). Walk over and talk to them often. Ask what problems they are seeing. What resources are tight? What won't management let them do that they think they need to do? Your project might be able to help them get some of that done. It could be a win-win. Or maybe just a not-lose/win.

There are subsets of operations types. There are the Unix system administrators, the database administrators, the network administrators, and—please do not forget—the help desk.

Help Desk

The help desk can make or break your system. The help desk is your first chance to change the perception of your system in the minds of the users who are having problems. I say having problems because they would not call the help desk if they weren't having problems. The ideal system user would never have a problem, but we live in the real world. In my experience, during the first few months of a system's life, users send a slew of problems to the help desk that have everything to do with change management and reveal no flaw in the system itself. The problem is that it's new, people aren't used to it, they get confused—possibly frustrated—and we can only hope that they call the help desk. Why? Because if they don't, they'll just slam the system to their friends, and it all starts to unravel. If they call the help desk, then you can fix it. So, if you get the help desk prepared for all the likely problems users might call about, you can nearly double the chances for system acceptance.

Now let's look at what the help desk is all about. We are led to believe they are there to support the users, but the help desk is seen by senior management as a necessary evil, just like the operations team. If those idiot developers would just make good systems, then our idiot users would be able to use them and we'd have no need for a help desk. Well, that's the thinking. So, the help desk is another function in the IT department that no one wants, but they at least acknowledge that it's needed. Help desks are so loved and treasured that they are frequently outsourced. They are not even considered a core competence to a company. Companies hire an outside firm to answer user telephone calls and try to keep things running. They sign contracts to try to incentivize the team to do better.

So how does management decide how well the help desk is doing?

They look at the number of calls—not a bad metric. Then they look at how quickly those calls are answered. Help desks use automated systems to log each call, record who answered it, categorize the call, and list the result. Management puts a lot of pressure on the help desk to resolve user problems in the first call. That's the primary metric management uses to evaluate the performance of the help desk. The unintended result is that the help desk is now incentivized to get the user to admit that the problem is resolved…even if it isn't. The help desk will close the ticket, get their bonus, and voila! Everything is right with the world. But the user still can't do whatever he or she needs to do.

Help desks have troubleshooting scripts that they use to help solve problems that have been seen and solved, at least the most common ones. We've probably all been walked through a set of questions that lead to a set of possible solutions, and that's what the help desk relies on. Most help desk staffers are *not* computer gods. They are not the most technically savvy people in the IT organization. They are people who are good at figuring out which of the troubleshooting scripts to run. If they can't solve it with a script, they will hand your problem off to the operations team, aka tier 2. Once they do that, they've met their performance goal, so there is a strong desire to get the user handed off to anyone as soon as possible to improve the metrics.

Thus, it is incumbent on you, the development team, to help the help desk help your users. Again, the help desk doesn't know your system from a hole in the ground, so your system can only mess up their performance unless you help them.

When do you talk to the help desk? Well, fortunately, not as soon as you need to contact almost everyone else. You can actually wait until you are getting ready to deploy. What should you do?

Ask them what they need. No one ever does this!

- Ask for *their* format for the troubleshooting scripts.

- Make a list of who the users will be.

- Make a list of who can grant permissions/access to the system.

- Make a list of the second- and third-tier points of contact.

- Make a list of the most obvious things that people could get wrong and the solutions.

- Make a list of the solutions to less obvious problems.

- Make a list of system dependencies so the help desk can do the initial debugging (e.g., I can't log in, check to see if Active Directory is up).

- Give the help desk a real demo. Help them understand the following:

  o How the system works

  o What the business is that it supports

  o How users could get bollixed up

  o Who owns and controls everything

  o Who is going to be calling with angry tones, etc.

- Introduce the development team, who are in essence your tier 3 help desk, to the tier 1 help desk. Explain whom to try first for specific types of problems (e.g., reporting, login, business logic).

- In other words, help the help desk to help you.

You want to leave them with the feeling that they can handle anything that is coming their way.

Remember all those stories of insanely stupid users? Those came through the help desk. You want them to think of you as having their back. You want them to think that you've prepared them for anything, *and* you want them to feel comfortable putting the user on hold, calling you, and resolving it without it ever appearing in the tier 2 or tier 3 pile. In other words, you want them to get complete credit for resolving the issue. You want an aura of confidence and a tone of assured capability when they are talking to the users about your system.

When something new comes up, immediately offer to update the troubleshooting script. Odds are, they will tell you that they'll do it themselves, but you have just scored major points and confirmed that you have their backs, that you understand what they go through. You must never forget that the help desk is the source of all those great Internet postings about idiot users. It was the help desk that got the call about the foot pedal (the mouse) and the cup holder (CD drive). They are the ones who get the calls that nothing is working when in fact the user has not turned on the machine. They live a tough life, and you should thank your lucky stars that they can answer these questions with a smile on their face and a lilt in their voice, because you sure couldn't.

When you launch, bring in a basket of treats for the help desk staff as a pre-thank you. Make them feel like part of your team...because they are. Stop by the desk a couple of times a day to find out what sorts of issues people are having and in the process find out how good your troubleshooting scripts are. Ask the help desk what else they need. Be there for them!

Unix System Administrators

If you've ever had a normal conversation with a Unix system administrator, you were mistaken and were actually talking to a homeless person. The attire, grooming, and social skills are very similar, so it's an honest mistake.

Unix system administrators are an interesting breed. System administrators have always been the keepers of arcane knowledge— they who keep the systems running—and frankly, if you've had a normal conversation with a system administrator, well, they probably weren't a very good one. They are some of the least communicative, least socially adept, of the engineers. Developers complain about the rudeness of system administrators, and that's saying something. They also appear to be the least helpful of the lot, but that's actually a misperception. They are very willing to help you but only if it fits within the ideal model they have for their environment. Again, let's look at the life of a system administrator.

Their job is to keep the servers running at maximum efficiency, and, like every other person in operations, they are beaten soundly when anything fails and ignored when everything is going aces. They have spent years learning all sorts of specific optimizations. To a system administrator, the ideal system is servicing its applications with no individual resource being overly consumed. By that I mean the system isn't waiting for data to come off the disk drives, it's not flooding the network with traffic, the CPUs are running at a reasonable rate (say under 70%), the memory the system uses is not excessive, and the swap files are small and lightly used.

System administrators are responsible for keeping all of the system software up to date, which is a major effort. On the average home PC you'll find Windows or maybe the Mac operating system, MS Office, Adobe Viewer, Firefox or Safari or IE browser, Norton or McAfee virus prevention software, some photo editing package, and other odds and sods. We've seen all the update notices that come up on the screen. Security patches, features upgrades, support for new hardware—all sorts of things. There are several every week. Sometimes when we apply the new browser update, some of the plug-ins don't work. A couple of days later you get the notice to update the plug-ins. In the meantime some functions in your browser seem to be squirrelly.

Well, now imagine that you have all of that to deal with and multiply it tenfold or maybe more. That's a typical server environment. There are libraries and libraries of code that other code uses to support other code. Language updates, security updates, database system software updates, web server updates, browser support, driver updates—you name it. The system administrators have to get these applied so that the systems are secure and up to date and supportable. But you're building software and testing things and trying to deliver an application. If they updated the system every time a patch came in, you'd be trying to build a system on a moving target. So really, you never want them to update the system...except your team is going to want to use that brand new Monty tool that is not currently on the system...so you want them to change the system right away to support your application...but not change anything else...even for that other development effort. You're talking out of both sides of your mouth, which the system administrator sees and reasonably decides you're completely unreasonable and clueless to boot.

Every update has to be tested, and testing things at the system level is tricky. Most business applications run in annual cycles. How do you effectively test a patch to an operating system for every application in the enterprise and for every set of functions it performs across the entire business cycle? The books tell you that you use automated test tools and perform complete regression testing on every application in the test environment before moving it to the production environment, but I'm here to tell you that 90% of the shops don't do this. They can't do it in the time frames they are given. It's just not doable. They test the core features that they know well, see if the system appears to be working, and give a thumbs up. Then when it breaks and we go from the help desk tier 1 to tier 2 to tier 3 and tier 3 (developers) are running it down and it used to work and now it doesn't and what changed and oh yeah, those jerks in system administration patched the system and why do they do that, they're so stupid...and so on.

So really, the users test the system, and when there's a problem it nearly always results in another round of shots fired between the Hatfields and the McCoys.

Let's get back to the Monty tool you want to use. To the system administrators, this is yet another new but fundamentally redundant thing that can go wrong, that requires patching, that you only want because it's pretty and shiny and the latest thing. We already have a tool that does what the Monty tool does. If every application would just use it, the system would be cleaner, easier to maintain, easier to debug, easier to please users, and cost less. The system administrators are not always wrong. Oftentimes development teams want to use a new tool or language because it's trendy. Rarely is a new tool suggested because it meets a brand new need. Developers want to stay current with technology, and who can blame them?

Here's a quick aside, when I get on my soapbox. New tools should be brought in when:

- Old tools are going out of business
- Old tools are too slow to support the business
- Old tools are flaky
- New tools can consolidate (easily) the functions of multiple old tools
- New tools will support *real* future needs that old tools don't. I say real because I've sat in rooms on multiple occasions when people were examining expensive new tools that do things

they will never do. I've heard people talk about moving to a new search engine because it can search photograph libraries. Great! They didn't have photograph libraries and honestly never would. But it was cool.

System administrators are not immune to the shiny new tool syndrome (SNTS). The administrator world is full of them. Although they don't often cost anything on the Unix side of the house, they are often pretty pricey on the Windows or mainframe world.

System administrators have to make backups of the system and the data on it. We've all deleted a file erroneously and had to take hat in hand and beg the operations team to please restore it from these magic backups. By the way, when they do, you need to thank them sincerely because it's a major pain in the butt to do that. Backups are an enormous resource drain, *and* although we are completely dependent on backups to recover from human oopsies and major system meltdowns, we don't like to be inconvenienced one iota to do them. So backups most often are made during the wee hours of the evening. Since we don't always have system administrators in the office in the wee hours of the evening, we have to have tools, scripts, and the like that the system can run on its own while we're sleeping. Those scripts first have to stop users from making changes to the system, so they have to either cut off access completely or limit it severely. Next they have to either make a complete copy of everything and anything on the server or find out only those few things that changed and just copy them. With no one around to help.

When system administrators arrive in the morning, the first thing they check is whether all the systems are up and running. They may have been awakened in the middle of the night by an automated distress call if the systems didn't come back up. The very next thing they check is what happened with the miraculous backup. If the backup didn't run properly, then the enterprise is at risk. We've lost the ability to recover from mistakes made yesterday or a failure in the system today. If that risk manifests itself, the system administrators get flogged. Doesn't matter if your application did something "cute" in the off hours that caused it; it will still fall on the system administrators' heads.

So you need to work with the system administrators to find the right time for your application to run batch processing or nightly tasks. If you don't, you're in for a world of hurt.

Most everything the system administrators need to do to prevent the servers from crashing during the day can't be done during the day. The nature of the beast is that they spend a lot of weekends and late nights in the office making sure your servers are there to support you. Lots of times things go wrong in the late hours, which turn into later hours. We sometimes perceive them as strolling in late in the morning (oh, so nice to see you!), looking ragged (fashion is also not a strong suit of the system administrator crowd) and sporting a really crappy attitude when they are asked to please reset some password right away and why isn't someone here when I come in at 5:30 AM so I can pick up my kid at 3:30 PM when I leave the office. Truly. They're coming in late because they were up all night making sure that you would be able to do your work in the morning. Most business users would expect a medal if they pulled an all-nighter. They'd expect some comp time and kudos from their managers. System administrators get zero extra dollars and no credit at all for this extra effort, so be good to your system administrators.

Left to themselves, the system administrators could fill every minute of the day designing, creating, and testing scripts and utilities to monitor and update the system. They would be happy as clams. Updating the operating system on a data center's worth of servers is an enormous effort. It takes weeks of planning. One has to schedule which servers will be updated in what order, figure out which applications are affected (and therefore which users), which development team, which set of testers, who's going to be there to test, etc. etc. And that's just in the test environment. You get to do it all over for real in the production environment, assuming everything goes well.

Every one of those applications has a business function it supports (well, it bloody well better have a business function!), and that business function has a business cycle, and at certain points in the cycle you can't just shut it down or make a change. Sometimes the whole system has a cycle; other times it's individual transactions that can't be interrupted. The system administrators need to coordinate with every user to ensure they know that on Friday between 4:30 PM and 7:30 PM their system won't be able to accept new orders. Everyone agrees and everyone forgets until 4:20 on Friday that Bob is working on the West Coast that day. So, the schedule is a mess again.

Mind you this is also complicated by the fact that the system administrators will send out notices that are indecipherable to the business community.

"TCP/IP traffic across port 80 will be unavailable between 2130GMT Friday until 0030GMT Saturday for applications running on Darth, Leia, and Han. Caching transactions will also not be supported during this time. If you have any questions please contact system administration."

Well, that's clear as mud. Who knows which servers host their application? No one. Who should have to? No one. Who knows what type of protocol is used to support their application? No one. What's a caching transaction? I dunno, no one knows. And who the heck remembers how many hours we are behind GMT? So the friction among the business world, the developer world, and the system administrators continues.

Database Administrators (DBAs)

There are at least two flavors (arguably three) of database administrators.

- Development DBAs
- Operations DBAs

The former is often part of the development team and is the one who helps create the tables to store an application's data and controls the operation of the system.

The latter is usually part of the operations team and keeps the database management system up and running, performs backups, and upgrades the database software.

Development DBAs are odd ducks in the engineering world. Many times you can speak to them...in English! Development DBAs want to create order from the chaos of business data. They want to identify the key business objects that we care about—invoices, payments, employees, divisions, products, and the like. They then want to fully describe each of these: invoices have customers and products, they have dates and quantities and prices and totals and account numbers, and other things. Developmental DBAs like to lock in the relationships of the business objects in the database structure itself to save the coding world from making mistakes. For example, they like to hard wire rules like "each invoice has one and only one customer," or "An employee has a last name, a middle name, a first name, and perhaps a Jr, Sr, III and that's it." They like strong, rigidly controlled sets of data so they can make accessing them, updating them, and maintaining them quick and easy and reliable. It allows them to be a star. Applications work quickly, nothing gets corrupted, life is good. James Martin proposed the notion that while business processes might change, the underlying data objects do not. No matter how we process an invoice, the data about an invoice are the same. So starting one's system from a cemented foundation in data relationships is the best thing one can do. Once you do this though, you really have a dilly of a time changing it. It is a nightmare!

Organizations set up whole management structures to review proposed changes to data: "Will we allow hyphenated names into our database?" The DBAs hate such changes because they might create problems with the code when it comes upon a special character like the hyphen. Maybe the field length won't support the name. "Merrill Lynch Pierce Fenner & Smith" just might exceed the space allotted—oh and notice that ampersand in there…what a pain. So, DBAs are magicians in that they can take the chaos of a real-world business, identify the discrete things we care about, neatly describe them in terms of the information we need to know about them, and then relate them all to one another based on the business rules. That is a remarkable skill really. It requires both a strong ability in abstract thinking *and* an attention to detail that are rarely present in the same person. They have to understand the *business and* then be able to model that in its technical representation. So cool. Actually very cool.

The problems crop up when the real world intrudes. Business rules are so often not rules, but norms. Ninety-nine percent of invoices probably go to one and only one customer. But sometimes the company sends the same invoice to each member of an LLC, say, the Mario Brothers. There are cultures in which people have multiple names: "John Jacob Jingleheimer Smith" (whose name is not mine too) or "George Herbert Walker Bush." There are people who go by their middle name—"H. Ross Perot"—or maybe by no name—"E. E.

Cummings." DBAs hate this. They will fight it. They will tell you to stuff all the names before the last name into one field or not to include the extra names. They will tell you to pick one of the Mario Brothers and only send the invoice to that one. Or they might tell you to create separate customers for each of the Mario Brothers and duplicate the invoice and order information. They'll tell you to start abbreviating names if they don't fit. They'll tell you that names can't have ampersands in them. They'll tell you all sorts of things that will protect the perfectly formed model of the data from being despoiled to preserve the perfect order they've created in the database.

In an instant, that amazing person who was able to hear the business process and make something technical out of it, speak with the humans in the firm, ask great questions, and show them things about their world that even they never appreciated turns into a petulant teenager who doesn't want to sit at the table for dinner. They get sulky; you've messed up their perfection and why? Because people do "stupid" things in the real world. "Honestly, I have to mess up this complete, normalized database because some idiot wants to send out the invoice to three different customers who are really the same customer? How does that make sense? Why would you do that? If you have customers like that, get rid of them! Just send it to one and let them figure it out. Just pick three of your names already. No one needs all those names. You can't combine customer numbers; they're unique. What would that mean? I don't care if the companies have merged; the data shouldn't."

What's also fun to watch is when one DBA takes over the work of another DBA who has a different slant on things, or even better yet, when they take over work done by an amateur DBA. My wife had to peel her DBA off the ceiling one day because he was genuinely enraged at the customer for naming a database column "LowIncome2+." LowIncome2+ has a special character, the plus sign. Unless every reference to this column name is properly delimited, the system will assume you are trying to add whatever comes after the plus sign to something called "LowIncome2," which, of course, doesn't exist. This is legitimately a problem. Sure, there is a way to handle it, but it is nearly guaranteeing obscure, difficult-to-figure-out defects later on.

The point is the DBA views this column name choice as imbecilic to the point of malicious. You must have had evil intent if you chose to do this. If it was a naïve move, it was unforgivably naïve and you must die. Most business users would probably just say "what?"

A major threat to the DBA world is indexing and search engines. Google is an example of this. They can't control all the data, and search engines don't require the data to be in a rigid format, and they only get the control data, and they hate you and your stupid system. Now the proletariat of the technical world can access data without them. This is not good to a DBA, not good at all. Search engines don't work with neat rows and columns of data. They take text in any form and index it. The concepts of fixed relations and foreign keys between tables are gone. They're not needed. So the DBA loses all control. So do the enterprise data management people and the standards people and a slew of other people. It will be interesting to see how this plays out in the coming years.

Maintenance DBAs are much more like system administrators as far as demeanor and priorities. They monitor the space used by each application's database, expand it when needed, and optimize the indices periodically to keep the system running quickly. They watch the performance of the system and analyze how the system is being queried to see if there's a search or function that takes up what they consider to be too many resources. They handle requests for access and, most important, handle the backups. Like the system administrators, if left to their own devices, they'd be spending the whole time writing scripts to identify problems, perform database clean-up, improve backup performance, and automate permission changes. They'd be thrilled to get the time to do that. Instead they spend an inordinate amount of time with operations coordinating update schedules for the next version of the operating system, restoring old copies of databases for applications to resurrect what went wrong with their application data, and performing all sorts of tasks other than achieving perfection in operational database performance.

That Third Type of DBA

Earlier I mentioned the third type of DBA. This is the enterprise data architect. EDAs are a whole 'nother type of critter. They are the next level up in terms of data order zealots. They don't just want data to be ordered and clean and pretty for each application; they want the entire enterprise's data to be ordered, clean, and pretty. They want one and only one set of customer data. If two applications need customer data, they should get it from that one and only one place. They want people to be identified the same way in all applications everywhere. For example, all applications should store names as first, middle, last name. No application should store a name as just one big field. They are the abstractors of the abstraction.

These EDAs live in a process sphere called enterprise data management. There are tools and processes dedicated to this. They try to pin down all of the data that are used in relational systems to link tables together. They try to standardize the look-up tables across the enterprise. It is a noble, albeit Tantalus-like exercise.

They rarely win. They are frustrated *all* the time. Each application team has a budget and a schedule, and if the choice is to meet the EDA's needs or their own deployment schedule, no one will get points from the division they support for meeting EDA's needs. It will be put on the list of things to do next deployment or, more likely, the "never" list.

EDAs don't get budgets. They are the unfunded preachers in a small town.

EDAs will talk your ear off. They will go to the CIO and tell him or her that your team is not cooperating. They will attend every meeting and take up time; they will, in short, be a complete nuisance. They're right, but they can only get you in trouble.

Throw the problem back on them. If they want you to use application X's data, pin it on the EDA to get you the interface specification. Make them ensure that the models are compatible. Budgeted value field looks good on the surface, but what does it really mean? Make them figure it out. Never buck the EDA. In many organizations they have gotten to meet with almost every division. They often have the ear of the CIO. Why? Because they are all about reuse—this mythical creature that, like Sasquatch, is so rarely seen and then only by people who are not themselves credible. Reuse is the Holy Grail to CIOs. Imagine paying to have something built and then being able to leverage that investment again and again! It would be fabulous! It just happens so rarely. With all the effort and attention devoted to reuse, it's too hard for organizations to swallow the argument that it's a good idea to delay deployment of application X by three months and add 25% to the cost (thereby blowing the ROI figures) to be able to support some unknown application's needs down the road…to be compliant with the religion of reuse.

It makes sense on paper. We in the USA just don't think long term like that. *And* I fear it's a myth anyway. By the time application Y comes along and needs this new generalized feature, the data, business rules, business timing, or something else will have changed. Also, interfaces don't always keep up to date with changes in the application. But CIOs have to find a way to do more with constricting budgets, so the shiny lure of reuse almost always fascinates them. In fairness, it's one of the few areas in which they can control the costs. Product licenses and maintenance are costs set by the vendors. The CIO can't affect that. They can't control the cost of electricity for the data center, air-conditioning repair and maintenance, the price of computers, prices for bandwidth. What can they control? The development budget, and now here comes the EDA talking about "buy once, use many times." That's such a relief!

Network Administrators

Good news and bad news on network administrators. The good news is that you rarely have to interact with network administrators. The bad news is that when you do, it's a problem. The network administrators are responsible for keeping the network up and running. They manage the switches that separate traffic on the internal network so that you don't wind up with one big party line and come to a slow crawl. They maintain the firewalls that keep all those nasty people from attacking your data. They manage the connection to the Internet that is provided by some large ISP. Personality wise, network administrators are some of the most reasonable members of the IT team. Still a little rough around the edges at times, they generally enjoy making the switch jump through flaming hoops and take pride in protecting the world from hackers via their firewall.

For most applications you won't have to do a thing with the network administrators. When you do, it will be either a simple problem or an impossible problem. Simple problems are things like you need a separate subnet to keep all of your application traffic separated from the rest of the company. It might be that your application is handling HIPA data or has employee privacy issues and you need to ensure that the only ones who can see it are the HR staffers. Subnets are not a problem for the network administrators. They'll need someone's approval (usually security's), and then it's just a fun little switch configuration task. They have no pressures to push back on this request. They will gladly help you.

Impossible problems come when you have an application that will exceed the capacity of the current infrastructure. I ran into this once when we were working on a system that was monitoring for regulatory compliance by scanning all blog posts for bad actors. We were pulling in mountains of data every night and then processing it all day. We exceeded the capacity of their network.

A great analogy was given to me by my friend Hong Kim. Imagine the corporate network is an inner city street. Lanes of traffic go each way, and there are sidewalks on both sides in front of buildings. Most requests that you make are doable because the network administrators can decide that we'll have three lanes of traffic go uptown and two lanes going downtown. They can decide and they can control whether or not there are turn lanes on certain blocks. They can even, in a pinch, narrow the sidewalks to create a new lane for traffic. But once you exceed the capacity of the traffic system, you are truly in trouble. You have to tear down the buildings on the left and right to create more lanes for traffic. Impossible issues that you bring to the network administrators require a complete rip-out and replace, and *that's* what makes them impossible problems. If you exceed the bandwidth of the Internet connection, or the switch, you have to completely replace them.

The network administrators are not opposed to doing this on principal. Bigger and more is better in their eyes. The problem is that it is an enormous task that might require months of planning and work, *and* it's very expensive. They probably have no resources for this in their budget. If your project was funded for such a contingency, then you're in much better shape, but odds are you weren't. These kinds of problems will require approvals at the top level of your organization and are worthy of a completely new project all by themselves. So if you get a jaw-dropping stare from your network administrator, you probably just raised a "raze the neighborhood and rebuild" type of problem. Before either of you panics, see that you understand the required bandwidth and concurrency. See if you can change the concept of operations so that you are spreading that load over a longer period of time or perhaps in short bursts. See if you can segment the work. Do not take on these sorts of tasks unless there's truly no way to avoid them.

The network administrators will have to take point on this sort of thing. It is not part of your project even though you are the driving requirement. You are changing their entire infrastructure, and they need to lead the effort. Odds are you do not have the proper perspective to head the project. You are focused on what your application needs, whereas they are responsible for every part of the enterprise's networking needs. You are likely to suboptimize the system to meet your needs and leave them with a nightmare to take care of the rest of the company. Let them handle it. Realize that you are now about 90% likely to hit a delay on your project because of the time it will take to create the proper infrastructure for you. Be supportive of your networking team as they make their project plan, get budgetary approval, and deal with every other project in the enterprise that will be affected by this major infrastructure change. Never take a shot at them for not having the foresight to imagine that a project like yours might come along and exceed their capacities. Even if they did, there's no ROI argument to put it in place before you come along, so it should never be approved in advance.

Testers

A lot of testers started out their careers slinging code. For whatever reasons they found that they most liked the part where you exercise the system to find out which demons need to be exorcised. Building wasn't the exciting part; it was validating and stressing that made them happy. This is a good thing because the more your testers know about writing code, the better they tend to be at breaking things. And breaking things is what testing is all about. The goal of testing isn't to make sure the system works; it is to find out where the system is broken. It is surely broken, but you just don't know where. That's what good testers find out for you.

Good testers are people who are very comfortable with charts and tables and cross-references and reports. They get to create a whole lot of them. They are very detail oriented and want to ensure that nothing slips through the cracks. Your tester is often the first person to tell you that you've missed some obscure requirement. Everyone else missed it, but not the tester. Testers like order, and they are frustrated by the chaos of the end of a release. If they've been around a while, they know it's coming, but they still don't like it. You need to be supportive of your testers when the crunch happens. They are on the front lines, and all of your desires to just push the system out to make the client happy are at odds with your tester's counsel to hold until it's of better quality. Usually you should not listen to your inner demon but listen instead to the tester angel.

Good testers are evil geniuses. They contrive amazing ways in which your system can fail and then run that scenario again and again, getting joy each time it crashes and burns. I don't think they get joy from watching your face as the system immolates itself, but sometimes I wonder. Either way, better they find it than your users, so the more evil your tester, the better your system will be.

Testers want to make sure everything is perfect with your system. It never will be. Instead you will come to a decision point on whether the imperfections are sufficiently bad to delay delivery or if it's okay to deploy the system with these flaws. The testers will hate this every time. Their passion for perfection is what makes them good at their job, yet the very nature of the business means that they will never achieve their goal. Sometimes they can feel very defeated after a release. It's incumbent on you to let them know that everything they did is appreciated and recognized and that you continue to admire their drive for perfection. Never for a second let them think you don't care about the quality of the system. There is nothing more disheartening to a tester than to believe that no one else wants to make the system the best system that's ever been assembled.

Many testers are process-oriented people. They like to start with the requirements or user stories and acceptance criteria, build out the test cases, build out the test harness, and then run the tests, record the results, and retest until it finally works. For a lot of this kind of work automated tools can reduce the effort required. Most testers jump at the chance to relieve the monotony of manual testing and instead spend more time devising evil approaches to breaking the system. If you find you're working with a tester who prefers to do it all manually, you need to break them of that habit as soon as possible. Make sure your testers know you recognize and appreciate their evil ways but also thank them every time they automate something, because the automated tests make it easier to process last-minute changes.

Information Security Managers

Imagine you're responsible for the inventory in a large warehouse in a sketchy neighborhood. There have been problems with thefts from the warehouse, and senior management has made it quite clear that it is your responsibility to shore things up ASAP. But this is an unusual warehouse. Lots of people work there, and they all have keys to every door. None of these employees reports to you. In fact, you hardly know any of them. You have sent emails on a regular basis to everyone, informing them of the need to make sure all the doors are locked or the company could be at risk again. You explained the impact on the company's bottom line. You put signs up next to every door, making it clear the door should be locked.

But the employees don't listen. They leave the doors unlocked because they think it's a nuisance to lock and unlock them. In fact, lots of times they take inventory home but don't lock it up, and it gets stolen from their car or backyard. To make matters worse, sometimes they tape the keys to the outside of the door so they don't have to remember to bring their keys, and thieves find the keys and steal the inventory. Management still comes down on you when the employees do this. They want to know how you're going to solve this problem.

As if that weren't bad enough, the thieves are continually coming up with more and cleverer ways to get into the warehouse—ways that no one has seen before or could have been predicted. They snuck in inside packing crates, so you put in a screening system for all incoming crates. They entered dressed as security guards, so you put in a new badge system. You're always one step behind the thieves. After each break-in management came down on you and blamed you. They think you're incompetent. They wonder why you haven't gotten these thefts under control.

An information systems security officer's (ISSO's) life is very similar to that, only much much worse. The warehouse manager doesn't have guys coming up with plans to create large new holes in every wall of the building to support automatic forklifts that are controlled via panels in the local mall food court. He doesn't have armies of people suddenly show up at every door clogging up the entire operation by each asking if Bob is there thousands of times a second.

By design, the ISSO can never win. If no data are ever lost or no one ever blocks operations, then he did what he was supposed to do. He breaks even. But anytime some bored teen with a couple of PCs decides to go after the company's website, it's the ISSO who's going to get blamed. It's a classic case of blaming the victim. We don't blame the police department when a home is broken into, but we sure blame the ISSO when there's a data leak. We even blame the ISSO when the leak happens because an employee left the "door" open. To the ISSO your project is just another automated forklift threatening his already shaky world and one more opportunity for him to get fired. There is no reason he should be pleased to see you and your team. But it gets still worse.

Security people…There was a time not that long ago when many ISSOs were administrative assistants who decided to improve their lot and got the job as the security person. This made some sense in the days when all secure materials were on pieces of paper and you needed someone with good organizational skills and an attention to detail. Many moved up in the ranks of the security hierarchy, some into the corporate chief security officer role. Then computers entered the picture, and the security people were ill prepared for managing them. Although this has changed some as the need for information security has become painfully evident, still many senior security officers are not technical. There are now industry standard certifications, and commercial tools, techniques, and best practices have improved, although odds are your security person won't know any of this. Currently less than half of all companies require an IT security person to be certified. Honestly, dealing with the security people has been the most frustrating experience in my professional life. The industry certifications should have weeded out the truly incompetent, but I know of cases in which the information security manager could not pass the certification exam so he hired minions who could. This was not a good situation.

Many security managers are angry, and I have some theories as to why. Some are frustrated that they don't truly know what they are doing. Many hate the software engineers telling them they're idiots (imagine that!). Most are underfunded and unsupported by management and overwhelmed by the spectrum of threats. No sooner do they figure out what to do with CD copies of corporate files than someone comes along with a proposal for cloud storage solutions. They are the ones stuck figuring out how to fully secure the enterprise's data while making the data available anywhere at any time. If you think about it, it's not a simple task. Every time the enterprise is hit with a denial of service attack, the yelling starts over at the network operations side but quickly gets refocused onto the information security team. "Why aren't you guys doing something to prevent this sort of thing!"

Telling senior management that truly there is no way for someone in a single corporation or government office to prevent an attack will get them nowhere. You can make it harder, you can do things to make it detectable sooner, but you can't prevent it. Users are constantly violating even the most obvious of security procedures. They'll download software onto their workstation from god knows where and then refuse to fess up when the building is hit with a malicious virus. Who gets yelled at instead? The information security guy. He's the one who has to find and remove the virus and then fix every file in the enterprise that might have been affected. People bring in their laptops from home and connect to their favorite sites, opening up a route into the network around the firewall. Some hacker will find this and access the organization's data. Who gets the blame? The information security guy. Oh, the users might get a reprimand, but the hammer falls on the information security office.

The IS department also has to figure out how to control physical copies of corporate data. Several large disclosures of credit card data came when a low-paid system administrator allowed some hackers to "borrow" backup tapes of the company's computers. Employees have been known to print out large swathes of sensitive corporate data to look at while at home and then lose them. Laptops, again loaded with corporate data, are often lost or stolen. Each time it happens the IS team gets a frownie face from senior management.

A further complication for the IS team is how to secure the backup infrastructure that more and more organizations are establishing to support operations in the event that the main data center is unavailable. These continuity of operations or failover sites are a nightmare. They are a completely different set of networks, computers, and copies of the corporate data in a different location that is not regularly used. So if you're a hacker, your entry into that system might not be noticed as easily as if you had to sneak around in the main network past a full-time network operations center and hundreds of users who might see that a file of theirs has changed or is locked by some other random user.

In response to frequent and unfair bashings by senior management, some security managers have adopted an "Absolutely Not" approach to life. Given their druthers, they would simply fire all the users and lock up the system to ensure safety of the data. If you're part of an organization that doesn't allow users to add any software to their machines, it's almost always because some numbskull installed something one day in the past that caused the information security guys to get a beating. If you aren't allowed to use the wireless features of your corporate laptop, it's because some dolt went to Starbucks on a telework day and carelessly opened up the corporate credit card information to the free world. Most people don't like to be beaten for the misdeeds of others. Given that the IS department is unable to stop users from doing stupid things no matter how often they are reminded or to make them take training classes, and no matter how many times users sign a piece of paper saying they understand the rules, IS has to remove temptation from in front of the users. The result is a strangled computer on everyone's desk and grumpy users who hate the IS department for doing this to them.

The security world is in response mode. If you have a good shop, they have the latest antivirus software on the network, servers, and individual PCs. They have set up redundant firewalls and keep them regularly updated. They have intrusion detection, network monitoring and alarming systems, and standards for enforcing strong passwords. This is the state of the practice, but keep in mind that every one of these layers is a response to someone who found a way past them. Antivirus software maintains a list of "signatures" for known viruses, but it can't easily detect a brand new virus. That's why your computer is always updating the virus files. The antivirus companies are extremely responsive once a new virus is found, but they aren't prescient. They can't detect a virus that they've never seen. Likewise, each time hackers devise a new way to attack a server, the network monitoring software gets patched to add this new method to the list of known patterns of attack.

The point is that all of the tools an IS officer can use are only as good as the previous attack. They are reacting to what's happened in the past. They can never know if what you're doing would present a new opportunity for a hacker to compromise the security of the enterprise. It certainly will not reduce the risk, and it's nearly guaranteed to increase it, so "No" is the default response.

At some point in their development, the IS department will institute a set of security requirements that no system can meet. With the intention of securing the enterprise's assets to the degree they are being told to by senior management, they lock down everything. I've been in situations in which the IS department told me that the data had to be encrypted on the disk and in transit from the disk or on the network. Both of those requests are very doable and almost standard now. They then went on to say that the data had to be encrypted on all user screens and printouts. Yes…encrypted on the screen. I thought I had misheard. I said, "You mean on the way to the browser?" He said, "No, on the screen." "But…then… the users won't be able to read it," I protested. "We've got to stop people taking screen shots to get around security," he replied. <Sound of head hitting desk>. In a similar situation an ISSO told me the data were never to be decrypted even during calculations. When I tried to explain that A times B equals C is not mathematically equivalent to Encrypted-A times Encrypted-B equals Encrypted-C, I was told I was out of my mind and it had to be. Again, these are examples of ISSOs who are not very technical. It's getting better but we're not there yet.

What do projects do when faced with this sort of thing? Usually you go up the management chain and get someone with authority to write you a waiver. The ISSOs hate that. They've been yelled at by senior management for leaks in systems that those same senior managers granted waivers for. They hate that, too.

Working with Your IS Department

With this as background, what's the best way to approach your information security team? The first step is to realize that they are trying to do right by the enterprise. They might be unpleasant (owing to constant beatings), but they have a nearly impossible job and you are going to make it more difficult by adding another possible set of vulnerabilities. As you did with the operations team, go in with the attitude that you are there to help them help you. Meet with the ISSO as soon as you can once the project is approved. Ask how they want to operate. Do they want to be in on all the review meetings or only a few? Regardless, invite them to any and all reviews. Make sure they know that your door is open to them if they can find the time to attend.

Ask what standards they adhere to regarding security. If they are reasonably mature, they will have chosen a set of standards appropriate for the enterprise. HIPAA, PCI DSS, CIP, GLBA, FERPA, and Orange Book are just a few. If they can't give you one because the IS department is not that mature or if they don't like any of them, ask what guidelines they need your project to follow. Sometimes they do have guidelines even if they have not adopted an industry standard. One client I worked with really didn't care about much except that all calls between services had to be done via HTTPS. They were adamant about that. That's fine—I knew what I had to do.

Ask the IS department what documentation they will need and at what level. Some will go with whatever you're already doing as far as the design documents go but then will need their custom questionnaire. Also fine. See if you can get that questionnaire from them in the initial meeting. You should get a feel for how hard that is going to be to accomplish. One questionnaire asked what CPU interrupts we would be invoking. Honestly, none of my team had a clue how to answer that. My literal engineers were then looking up what interrupts every browser feature might set and writing them all down. I

> If your security person is helpful, friendly, and knowledgeable, then you should make a donation to your favorite charity because you are in Fat City, my friend, and you need to restore some karmic balance to the world.

took the question back to the IS shop, and it turned out they had no idea what that meant either. They'd gotten the questionnaire from a buddy who worked at a place that did a lot of embedded systems.

Ask what scanning they require before a system can deploy. Sometimes they are the ones who run the scans; other times, they'll have you run it and then review the results. Either way is workable, but you have to find this out early.

Some IS shops truly want no involvement until the system is ready to deploy, which is a high-risk approach for your project. Most project managers have hit the security road block at the eleventh hour. If your IS department tells you this, add it immediately to your list of risks and make it an active task. You need to get them engaged no later than the early design phase if you're following any type of waterfall, or in an early sprint if you're going Agile. You have to know what architectural choices are forbidden or going to cost you later on when it comes to getting security approvals.

I strongly encourage you to find out how your IS person became an IS person.

- If they came up the administrative assistant route, they can be a bit rough around the edges. They spent years getting dumped on by managers and others as the lowest person on the totem pole. Then they went to school at night to learn this new skill, but if they're still in the same organization, they've not gotten the respect they deserve for this new position because people still treat them as the admin assistant. You should take pains not to follow this pattern and to give them the respect their hard work has earned. Phrases like, "Well, this is your domain, so I want to make sure I am compliant" or "I want to follow the guidance you've put out there" will go a long way toward helping the IS person know you're not going to follow the pattern of belittling.

- If the IS person is the failed programmer, they often have a certain resentment or distrust of current programmers. They want to do code inspections and get into weeds that they may or may not truly understand while maintaining a certain sneer and distrust throughout. With this person, a good approach is to ask them up front to list their greatest concerns and to explain how they'd like to go about addressing them. Make them define their criteria (nicely), and then see what you can do to meet their inspection requirements.

- If you have a security zealot, it's tough because they want everything nailed down below ground level so no one can ever access it. With them, you need to find out which security standard they intend to follow and then read up on that standard to find the proper buzz words and criteria so your system can someday actually be used by real business users.

- If you have a power person, recognize that they need to believe that they are in control. Work with them to find their schedule, their process. Keep echoing back that you've included this review or that inspection at their request/demand. Make sure that you keep telling them you are doing their bidding. They don't want anyone bucking their system.

Refereeing Software Engineers and the IS Department

Engineers hate the security people. The IS department's requirements only make what the software engineers are doing harder or, at best, more tedious. Encrypting the data takes time. It makes the engineers' system look slow. The users don't care if things are taking longer because of the security requirements; they are just plain unhappy. Security requirements deny the developers their moment of glory. To be fair, if software engineers designed security into their applications from the very beginning, it would be more efficient and not nearly as hard as retrofitting it later. It's tedious to modify every bit of code that passes data to ensure that it's being passed in encrypted form. I've had guys tell me they'll go back and fix that later, but they don't want the system to be slow at this point in the development. To the software engineer, the challenge and fun is in making the system dance. Designing security into the system is not dancing, but rather trudging. Including the security requirements is like trying to dance with ski boots on.

If the security team is not technical, you will have to make sure to be at every meeting the security team attends. The software team is already not happy to have to meet the IS department's mandates, but if they find out they're coming from folks who can't code their way out of a paper bag, then bad things will happen. They will, at some point, say something insanely insulting, and all your hard work to get the IS department on your side will go down the drain.

Before the meeting, you should explain what the security folks are up against and what their role is in the enterprise. Let the software designers know that it's the IS department that takes the hit when your project screws up. So yeah, they want to make sure *we* don't screw up. Remind them that we need them on our side. Woo them. Forgive them for their lack of technical expertise. When IS says something that's incorrect or naïve, don't jump down their throats. Instead, share the standard that you are following, or *gently* make your point using simple terms. Slowly, without sneering or condescension. Make sure you stomp on any misbehavior quickly and in a manner that is supportive of the IS department. Translate for the team. "Bob, I'm not sure I understood all of that. Are you saying that complying with this standard that the IS department is requiring would make no difference to the total security posture? How can that be? Is that because no other system is doing it, and therefore the data are already vulnerable, or because of some other weakness in the overall architecture?" Make your guy gently explain it.

Software developers will rail against security constraints of all sorts, even those that actually add value. They also are very rule oriented. If yours is the only system that has to meet this odious standard for security and the data are otherwise unprotected in related systems, then your team will feel picked on. If you're in that situation, turn to the security department person and ask, "I assume the plan is to retrofit all the other systems over some period of time? So eventually we'll have better protected the data rather than multiplying the instances with this same weakness? Otherwise you guys wouldn't be asking us to do this. Is that correct?" Now it's fair. All systems have to comply. You can even reinforce it with, "Okay, we sure don't want to do anything stupid and repeat security errors of the past."

Be especially careful when it comes to reviewing the security scan data with your guys. The software engineers often have a deeper understanding of what is and what is not part of your system and what is infrastructure or other legacy applications. The scans don't make this clear at all, and the IS department often will ascribe a problem to your application that is not actually within your control. Your engineers don't like being accused of doing wrong when it's something that's been out there for ages. In their minds, if the IS department were any good at all, it would have found and fixed that security hole long ago. Having it now pinned on this new application is unfair and beneath scorn. Once again, step in hard and quick. "So Bob, that service is part of the Oracle environment and not part of ours? Is that what you're saying? Do you know if it's always had this problem?" Now Bob can demonstrate his knowledge of the Oracle service, and security can go crab at the DBA team about the security hole. If you don't jump in quickly, you will likely hear something helpful like, "I'm sorry, what? Good of you to finally notice. That's Ora_write. Ora_Write has a problem not us. Do you know anything at all about how things work here? Are you kidding me?" Actually that's not helpful at all; it's just the opposite.

To head off altercations like this, you might even want to have your guys brief the security posture of the system during an early review. Ask them to keep it painfully simple. "We encrypt all calls to the database and the results that we get back from the database. We do that through the standard database function, not anything we're writing…" This helps the engineers and the IS department develop a common vocabulary for the components of the system. It helps everyone understand where the system boundaries are and are not. You can pre-educate the IS department in how your application interacts with all of the parts of the world that the IS department cares about. Then when it comes time to approve things, they have an understanding of your application.

Building Resilience into Your Schedule

One of my hobbies is woodworking. There's an expression that in woodworking you have a plan up until you cut that first piece of wood. From then on you're adjusting to every successive cut. The same can be said of every software development plan. So much effort and thought and creativity go into the plan…and then reality intrudes. Since every development task duration is a best guess, you are always finishing one early, the next late, and finding out that something is much more complicated than you thought. Sadly, hardly ever do you find something to be *simpler* than you guessed. Some of that is because the engineers are always thinking about the hardest possible task, or the ugliest situation, since they never want to fail, and vagaries lead to failing to meet the schedule. They try to stretch out the time allowed to account for this possibility, so you are constantly adjusting. People get sick, businesses have crises that pull their folks from your project, users aren't available to be interviewed, vendors release new versions of things, and...surprise surprise! sometimes they don't work.

One of my key guys was out for four months when one of his sons fell out of a third-story window (miraculously, he not only lived but is perfectly fine now). Another guy developed a brain tumor. A woman on my team ran into complications with her pregnancy and had to go on full bed rest three months earlier than she planned. I had folks get unexpectedly deployed with their National Guard unit. Other folks have had to leave for weeks at a time because a parent died, got sick, or got arrested and had to be bailed out. I've had people miss time because basements flooded, kids got sick, they got divorced or suffered mental breakdowns—you name it.

I've had a customer shut down because the federal budget wasn't passed. Other customers got acquired in mergers and had to disappear for months while they handled the transition. I've had customers with all the same sorts of problems that my team had. Occasionally customers were reorganized into new units and their replacements had a completely different vision for the effort.

We work in the real world, and in the real world things happen that are not within our control. Your plan should not be built assuming everything can be controlled. Instead, you need to build a plan that is flexible, that can adjust, and that can be resilient to real world changes. How can you do that?

Let me acknowledge that on first blush, building such a plan would appear to be completely anathema to what I just discussed in the previous sections in which I described a highly detailed, dependency-laden, and fully resourced straight-line project schedule from beginning to end. You need that pristine and blessed schedule to reach consensus on what is to be done and who is responsible for what. You need the approval on the associated budget. That is what I described, but as I just pointed out, your project won't go that way. Thus, you need to do more thinking and planning so that your plan stays closer to the plan.

Earlier I voiced my disagreement with the concept of risk management as a separable part of running a project. What I'm talking about here is different from risk management. Building a resilient project means taking conscious steps to make your project better able to absorb changes that might happen, good and bad. It's the difference between building a car strong and rigid enough to survive hitting a pothole and building a car with shock absorbers that can dampen the impact. *Risk management*, in my opinion, is more focused on ensuring there are never any potholes and replacing the cracked axle when you do hit one. *Resiliency* is all about acknowledging that you might hit a pothole; you might hit some debris in the road or have to dodge a spooked deer. There are indeed road hazards. We might hit some. We don't know which ones, so let's build a car that can handle hazards.

Resilience in Staffing

If you were lucky, you got to pick your team and had specific ideas about their strengths and weaknesses and how to use them best on your project. You might have lobbied hard to get an individual because he or she was the absolute best at one particular technology you needed for the project. If it's a new or rare skill, you might not have anyone else on the project who knows it. Although you thought you were reducing your risk, in some ways you just pegged the needle. If anything happens to affect the availability of this person, you are in deep trouble. Getting the expert is a great thing, but you don't want to stop there. You need to immediately start that expert mentoring someone else on the team. Even when nothing goes wrong, it allows your expert to focus on the big ticket items while the apprentice handles more mundane tasks. Cross-training is something that is usually dropped the minute we get into the maelstrom of development, but the consequences are that no one knows anything about the overall system, only their part of it. This will not help you. Code walkthroughs are usually dropped when time gets tight, but done right, they are actually one of the best mechanisms for teaching new skills and sharing knowledge about how the system works.

Do not chop up the system too precisely when making assignments. Intentionally interject some crossover on tasks. Too often we divide the system along lines that are very clear and understandable and that give the engineers license to solve problems in whatever way they choose. For example, in a three-tier architecture you might have one engineer do the database layer, another do the user interface, and someone else handle the business logic. This is generally a good approach assuming the work loads are similar, but even then you should work some overlap and cross-tasking into the mix. It is mildly less efficient, but it means that you have at least two people who've had their noses embedded in that section of the system and can either take over or help someone else take over if necessary.

You also have to be constantly looking outside the team for resources, even if you currently have no problems with staffing. As a rule, once you get your team assigned, everyone stops looking for additional members, and although this plan seems very reasonable, it leaves you completely vulnerable to staff outages. Keep reviewing resumes of others internal to the company. See if you can pull someone in for a few hours for a review or a code walkthrough. You'd be amazed at how much a person retains of the style of development, the overall goal of the system, the priorities, and the people on the project by attending a review. When something happens, you now have someone who is not coming in cold.

Another approach that can work is to reduce the amount of code that is based on a unique or rare skill set. Limit your risk by limiting what the new tool does, or how much of the system needs to know anything about it. If you can isolate the rare skill to a couple of services that every other portion of the system can leverage, then you have reduced your requirement for that rare engineer and hence your risk.

You also want to look at your team a little differently when considering resilience. Many times you're better off with a flexible journeyman than a gifted specialist. Too often we value depth of expertise in a technology over speed in learning new things. If your team members can come up to speed quickly in something new, then every one of them will be able to plug in to multiple tasks on your project. They might not be as fast as that expert, but you can use them in multiple capacities, unlike your expert, who is a one-trick pony.

As a quick aside, when you are in absolute crisis/crunch mode, you do not want to follow the plan I just described. If you are weeks behind and the villagers are gathering around the castle with their pitchforks and torches, you will almost always do better by pulling in the very expensive expert to perform their magic in their area of expertise and then leave. It will actually save you money.

Resilience in Infrastructure

Most medium-to-large organizations have a development environment, a test environment, and an operational environment. You move your code from one to the next as it is completed and proven stable. That said, the development environment in many organizations is, shall we say, asthmatic. Often the dev lab gets the old production servers when those are upgraded, and often many projects share them. The development lab is rarely kept as shiny and fast as the operational environment, and I can understand how management makes this call. That said, if you are working on a project on a wheezy, unreliable server, you are at risk for all sorts of badness. Recall that developers often have fairly free access to do things on the development server. You can be blasted out of existence by a careless act from any one of your developers or developers from any other project. You can recover much of your past work from a backup, but you lose productive time until the backup is restored, and you lose anything done since the backup was taken. It's more than just the day's work. You can also lose time when operations needs to upgrade the development server. If they need to patch it and the patching goes poorly, you can lose days.

The network administrators may need to take things down to create a new subnet or move the development lab to a new switch. It is rare that this happens without at least some machine or service suddenly becoming inaccessible. Database maintenance and upgrades can also stop all forward progress. One of my projects came to a grinding halt when a key license expired and the client and the vendor were having a battle royale over the fee for the next year.

To be able to keep making progress when your infrastructure fails, you need to take some actions. First, you should have a copy of your entire development environment on at least one of your developer's workstations. If the dev lab fails, then you can at least all connect to one of the workstations and keep moving. It will be slow, but slow progress is better than no progress. With that, you need a copy of your code repository. Every developer should maintain the code they're working on in their own local store, but the team needs the official version, and the disaster repositories need to be easily accessible as well. Licensing can be an issue for backup environments, but I have found that most vendors are willing to provide a second copy of a license for emergency use only with either no or minimal fee. Compared with the cost of having your development team sit on their thumbs, it's cheap insurance.

A common infrastructure delay is licensing. Your project is expecting to use product A, but right now you don't have licenses for product A. Perhaps purchasing is in negotiation with the vendor. Perhaps there are capital budget issues. Sometimes management is trying to make an enterprise-wide deal for a whole suite of products with product A as just one part of it. Regardless, you need product A and it's not here. One approach to avoid this infrastructure failing is to start out with open-source solutions. They are free and you can get rolling. I have tried this (for example, using MySQL instead of Oracle), but it hasn't always worked well. We had to go back and modify the code to meet the specific protocols of the licensed product when it finally came on line. I offer open-source software as a potential solution. If you can stick with the open-source solution once in the actual production environment, you're much better off.

The network is also something that is easily made resilient, though sometimes with the chagrin of the network engineers. A simple $50 switch and some cables will give you your own network connecting your development team. I once had a situation in which the corporate active directory service was in constant disarray. We set up our own network, as above, and installed a simple LDAP server on our renegade network. This allowed us to keep developing while the network folks and operations got through their travails with Active Directory.

Resilience in Process

When you started the project, there was a particular approach that everyone in the organization believed you'd be following. It might have been the traditional waterfall, it might have been an iterative development cycle, or it might be one of the variants of Agile. Regardless, things can happen that make it clear you should change your methodology. There's nothing wrong with changing how you're going to deliver things. What you deliver is almost always more important than how you deliver it.

I worked with a client who was a strong believer in the traditional waterfall methodology. They had a voluminous process document that was considered the bible of the organization, and they'd had success using this approach. However, when the top management changed, the new executive wanted to please a frustrated board of directors by showing some early victories. He needed to be that new sheriff who cleaned up the streets and brought law and order to the town. He didn't have a lot of options but saw an opportunity in "shaking up" the IT shop. As in many enterprises, the business users believed that the IT team delivered things on a glacial timeline and were unresponsive to their needs (for the record, they might have been right). He wanted to show that he could get things delivered faster. I'm not sure he cared much for what got delivered, but he had to show something pronto and the hammer came down on IT.

What a waterfall gains in robustness it gives up in speed. Continuing to try to deliver faster using that methodology was just not going to happen. We switched to an approach similar to SCRUM but with an iterative seasoning and were able to start deploying small things quickly.

Another organization had a very formal change management process with a change review board, forms for submitting requested changes, prereviews before requests went to the board, and the like. It worked fine for them until three of the members were assigned to an effort with national security implications and work came to a standstill. So instead, we went asynchronous and did everything via emails. This stayed true to the spirit of the process but eliminated the meetings that key board members just could not attend.

One client had a fixed deployment approach that required that the security review be done once the software had completed user acceptance testing, that is, just before we went live. I had been burned by this twice before with surprise security requirements or steps to be completed. Since the definition of insanity is repeating the same thing but hoping for different results, I changed the process. The IS office doesn't like last-minute reviews any more than we do, and they are truly not trying to stop progress. They have their mission to protect the enterprise's data and every system necessarily poses a risk. At the same time, I could not get my schedule approved if it deviated from the approved process.

After approval I spoke with the security team, and we laid out what they considered attributes of an acceptable architecture, what tools would be acceptable, and even a start on how we could validate this. We provided some documentation on what we planned to build, and, more important, we agreed that we only needed to review exceptions to this architecture and we could do that as these were identified; we didn't need to wait until the end. We simplified their workload and reduced the surprises at the end of the project. We still held our final review meeting right before deployment, but it was a pro forma session to assure senior management that the security aspects of the system had been thoroughly reviewed and assessed.

For some of my projects, the contract said we should prepare detailed requirements documentation and review them with the business users. The problem was that the business users had a terrible time understanding the documents and how they related to their needs for a new system. They were a very visual group, so we switched gears. We still produced the required documents, although with significantly less spit and polish, but instead of reviewing the documents, we reviewed mockups of potential screens with the users. We had one person in the room take notes on which requirements were "blessed" as we looked at the mocked-up functions and which ones required more clarity. It was the same amount of effort to go this way than to produce draft after draft of the requirements document. You have to keep adapting to keep moving forward, and my point is that nothing is sacred on the process side.

Resilience in Functionality

There are many reasons why individual items of business functionality can get stuck on a project. Often the business owner is unavailable for reviews or to clarify requirements or grant approvals and acceptance. Sometimes the problem is just a lot harder than was thought and will take longer. Sometimes the schedule said it was due now, but the client is focusing on some completely different part of the system. Regardless, you need to do two seemingly conflicting things: keep moving forward and constantly adapt to the situation.

What this means is that when a task starts sticking, you should look at it and decide if you should do one of the following:

- Skip it for now and pick it up again later when the source of the stalling is no longer a problem. Instead use this time to move on to something else more efficiently.

- Simplify the function to at least that part on which there is no controversy or ambiguity. Or,

- Just keep banging on it. If the task has been stalled because the technology is fighting back, then perhaps adding some time to the schedule for it (via your risk task) will be the answer.

But the one thing you don't want to do is to keep trying to move forward blindly on the same path. You have to have the mindset that the order of tasks in the schedule is somewhat flexible, and you should use every bit of that flexibility to deliver as much as you can as soon as you can.

One of the biggest obstacles to flexibility in a schedule is setting it up so that you have one large delivery at the end. The Big Bang schedule means that everything must be done on one and only one date. Following this path nearly eliminates any ability to recover from individual hiccups along the way. If you can, set up a schedule that has lots of smaller deliveries of functionality. That way if an individual function is stalled, for whatever reason, you can instead pull something from a later delivery into this one and slide the stalled one into that later release. Or you can deliver a strawman version with only rudimentary functionality in this release and deliver the final one later when the users are able to focus on your questions or the engineer is back from sick leave or the server is available again or whatever else was causing the problem.

Do not fall into the trap of thinking that because it's written on an approved schedule it *must* be done in that and only that order. The goal is getting to the end of the project with all the approved features completed. The specific order of completion is really not important.

Tracking and Managing the Project

The primary purpose of the schedule is to give your client something to beat you with. Its secondary purpose is to communicate to everyone what you plan to do, when you plan to do it, and in which order. The tertiary purpose is to show everyone how well you're doing against that plan. Sadly, for some managers, the fourth purpose is to lose oneself in MS Project. You must avoid the fourth purpose!

Managing the schedule means finding out where you are relative to where you should be and adjusting accordingly. So the first step is to find out where you are. That is not as simple as you might think. Some tasks are easy to figure out. "User Guide First Draft" is complete when I'm holding a draft of the User Guide. If my system has fifteen functions and the technical writer has completed a draft of ten of them, it's about two thirds done. Other tasks are much harder. "User Authentication Check" might be a function that accesses Active Directory and returns a yes or no based on whether users are allowed to do the function they are trying to do. When it's been moved to integration testing, it's done, but what about any state before that?

Engineers are legendary for the 80% complete task. Ask an engineer how close a task is to being done, and first you'll get, "Well, it's mostly done." Good news but not actually useful. What does "mostly done" mean? If you ask for a percentage, you usually get the 80% response. Next time you ask they know it has to be more, so you get 90%. You may then be on a path like the fabled grasshopper that keeps jumping halfway to the finish without ever getting there. If you don't set some sort of criteria, asking an engineer for a percentage completion on a task is usually exercising the engineer's facility with ratios, something they are very good at. If the task was four days long and this is day three, they are 75% done. What's the likelihood that they truly *are* 75% done, given the uncertainty of the estimation process and the chaos that has gripped an average project at this point? Not good.

If you want a reasonably accurate status update, you have to do two things:

1. You have to create a safe environment in which to report status that might be less than the elapsed time ratio, and

2. You have to remove subjectiveness from the reporting.

Creating a Safe Environment

This is the harder of the two. A status report is a public declaration of either accomplishment (Yay! Glory to the engineer) or inability to work to the schedule (Boo! Abject failure). Hit the target and engineers are heroes. Miss the mark and they are undermined completely. They have failed. They certainly don't want to fail in front of the entire team, but that's what they have to admit to with every morning's standup meeting. This is not good for the ego. So there is a self-protective tendency to hide problems and overstate progress.

One simple approach is to change how you ask for information. You could ask, "Bob, how are you doing on your task?" but that leaves Bob nowhere to go. It's all on him. He might not have come up with the estimate, but now he's a failure if he can't make it. Instead say, "Bob, do you think the estimate for this task is holding up? Is it enough time based on where you see yourself now?"

You should have set the stage by making it clear that the schedule is a lovely cake made up of layers of assumptions and guesses. You need to tell the engineers early on, before we start coding, that we need to try our best to hit the schedule but that it might not be doable. If it's not doable, they need to let you know as soon as they think there's an issue. It gets back to my "no surprise is a good surprise" mantra.

How you respond early in the project to someone who's behind will set the tone for the rest of the project. Make that person feel stupid, inept, or otherwise bad, and no one will fess up when you need them to. The first time someone is going to be late and tells you, swallow your panic and thank them for letting you know.

> Engineer: "I don't think I'm going to be done by Thursday."

> Manager: "Okay, thank you for the heads up. How much time do you think you'll need?"

> Engineer: "I think I can get it done by Monday COB."

> Manager: "All right. I'll figure out what that means for stuff down the road. Is this more complicated than we thought or just bigger than we thought? Are there any problems or obstacles you could use some support on?"

> Engineer: "Naw, it's just taking longer."

> Manager: "Okay, but let me know if you do find anything we could support you on."

Notice you don't ask, "Do you need help?" That's a gut shot to most people, engineers in particular. Asking if someone could use support is better. It's okay to ask for support, whereas asking for help implies a failure. Support implies a sense of teamwork. We are not Lone Wolf McQuade doing this on our own; we're here to get the job done, and we'll back folks up if they need it. Support implies that the engineer is still in charge, that the engineer has responsibility, and that others are handling "other" tasks around the central problem.

If someone now pipes up and says they think they'll be done early and could help that engineer, try not to start crying with joy, but you're there. The team is self-repairing.

Removing Subjectivity

One of the difficulties in assessing where someone is with a task is that this task has not been done before; it's a brand new piece of work. In your average piece of code, the majority of the effort is in trapping and responding to errors. How many possible errors are there in a piece of code? It's very hard to say. How many responses are there to each error? There could be a lot. So trying to come up with a definitive statement of code completeness is not an exact science. It is possible to make gross generalizations about the nature of the work and decide, at a high level, what the percentage of the total effort will be for that piece.

On the one hand, for a piece of code that is largely controlling the flow of operations, the majority of the work is in ensuring that you have the desired control logic working. So roughly you could say that you were progressing as follows:

> 10% done when the unit is coded
>
> 30% done when the unit has all of the required branch calls stubbed out
>
> 75% done when the logic flow is correct
>
> 90% done when the unit has all of the branches working with real calls
>
> 100% done when the unit is catching errors from those calls and resetting things appropriately

On the other hand, a piece of code that is responsible for providing data from the database would have most of its effort expended in ensuring the correctness of the SQL calls to the database. That code's completeness profile might look more like the following:

20% done when the unit is coded

60% done when the SQL is correct for valid inputs and correct data records

80% done when the code is trapping errors from the database

100% done when the code is safe from all manner of SQL injection

The exact steps and percentages are not important. The point is that an engineer can look at the task and tell you rough percentages of object states at which one can claim completion. This has to be done up front; it can be as simple as an email right before the engineer starts giving you the profile. You can then include that in the notes field in MS Project, and now there's no guessing. You've made it simple to determine whether you're done. Is the coding completed, yes or no? Yes, then you're 20% done. No, then you're 0% done. There is no 15% done.

Understanding the Status Information

That's the status side of managing the schedule. You next have to decide what the information tells you. On day one, the status information is not a trend line but a data point that tells you nearly nothing. If you had a task that was on the critical path and it didn't get done, then you are in trouble. But if that did not happen, then you have some status about the task. Over time the data can become very useful. Most projects have a reasonably straight line when it comes to difficulty, by which I mean that most of the work is uniformly complex over the duration of the project. You don't often start on the hardest stuff and then finish off with some trivial work. Ramping up and getting "in-stride" takes a while, but after that you're just slaying the next dragon, moving ahead, meeting the next dragon, and so on. The reason this is important is that you need to look constantly at how well you're currently doing against the average estimate and determine what that means to your schedule and budget by the time you expect to hit the end of the project.

The hardest part of this (since it's simply math) is being honest with yourself. If it's taken the team one and a half times the predicted time and effort to complete the first 25% of the project, then odds are good you've underestimated this job by 50%. Project managers often kid themselves that this underestimation is due to the startup phase and people had to get used to the project and the technical learning curve but now that we're over that, the pace of completion will pick up. Nope. Not gonna happen. Or at least it's extremely unlikely.

Imagine what would have to change to allow you to get back to a pace that matches the estimate. Right now they are cranking out work at a ratio of 1.5 actual to every 1 predicted. The team would have to crank out code in 33% less time. How could that happen? Wouldn't it be nearly miraculous if they could do that? They might be good, but that's a huge increase in productivity. Add the fact that to finish on time and on the original budget, the time already lost must be made up. With 75% of the project left, that means that starting today, the team needs to nearly double their productivity. For those of you who love numbers, the math works out to moving from a 1.5-to-1 productivity ratio to an amazing 0.83-to-1 ratio.

Never lie to the engineers and don't lie to yourself either. The project is

> **Earned Value Management and Custom Software Projects**
>
> If you were paying attention, you probably noticed that the productivity ratio calculation and the task completion approach are both concepts from Earned Value Management (EVM). There's a growing school of thought that EVM is the solution to software projects going astray. As practiced by most organizations, EVM is an extremely labor-intensive process that involves expensive EVM-specific management tools only usable by someone with a week-long training class. In my opinion, the most valuable part of EVM is knowing what your productivity ratio is, because with that you can estimate your completion. Those you can calculate yourself, so I'm not a fan of going all-out on an EVM approach.

not going to make that original budget and estimate. You must act.

One of the things you must do is not freak out the team. They are smart. They've seen the schedule (you made them look at it), and they know they're behind. They too can do math, and they know they can't improve their production to that great an extent. Don't assume they haven't noticed. It's probably worrying them something fierce. They can lose confidence in their abilities, so be open about it.

"Look guys, we took longer to get this first chunk of the project done than we planned. The plan was based on our best guess of what it would take to do the work, and we guessed wrong. You guys are good, but given that we would now have to double our productivity to meet the plan, I'm going to have to go back to the business owners and find out how they want to handle this. They might drop some of the requirements, or they may give us some more time and budget to do things. In the end, it will be their call."

But also keep them sharp.

"In the meantime, I need you all to do whatever you can to hit the remaining estimates. By that I mean the task durations, since the dates are already going to shift. That said, if you think we've underestimated some of those future tasks, let me know as soon as you can so I can roll that into the discussions with the client. Relook at those estimates and let me know if, based on our work to date, you think they should change."

And don't forget to thank them.

"I know we're off to an inauspicious start, but I want to thank you guys for all your hard work to date. There was a lot to do to get us up and rolling, and you did a great job on that. Now we have to see it through to completion."

Options

Management texts talk about the three sides of the project triangle: quality, cost, and time. They represent the three degrees of freedom you have in a project. You can add time to a project, you can reduce functionality or quality, or you can add money.

Most business owners do not want to change the functionality, budget, or schedule. They want their whole system on time and on the original budget. This is an understandable desire but may be totally unrealistic. In some organizations, the clients slammed their fist on the table and told the team to make those deadlines or else. A beaten down manager will, early in his or her career, cave to this chest beating and try to make that happen. The only way it *can* happen is if the team works uncompensated hours, that is, long days, weekends, and into the night.

You can make up a little time that way, but as I've said, there's a tipping point after about the third week of crazy hours when productivity actually dips below what it was before you started the crunch. I've worked in places in which crunches lasted months. The workers were proud of it, but what I don't think they noticed was that they were not very productive during the crunch. Even when it was over, when they were rested and could work normal hours, their productivity remained low. They had a mental expectation of what could be done in a sixteen-hour day, and if they hit half of that in eight, they thought that was great; they thought they were wizards. So you can't just add hours and get the schedule, cost, and functionality for free. You can get about three weeks of the schedule back at most, but then you have a tired and worn down team. By then you had better be close to a stopping point or you'll fry them completely.

Looking back at the management triangle of quality (read functionality), cost, and time, one could naïvely think that this is a simple tradeoff, but these three aspects of the project are not independent of one another. If you add time to a project, odds are you also have to add money. How else are you going to pay the team for that extra time? You can add money with the notion that you're adding labor by doing so, but there will be a learning curve to overcome and that might also require time. It will take time from the new staff and additional time from the existing staff, so it's time and money and in fact, more money than you are originally thinking. Also, that new person isn't often sitting on a bench just waiting to be assigned to your project. You will have to spend time finding and then on-boarding that person. Quality/functionality is not as simple as it might seem either. If you make a function simpler by 50%, you don't necessarily gain 50% of the time back. There will likely be modifications to the existing work to skinny it down 50%. You might only net 40% at best because the inner workings of the code are all connected.

As the project manager, it's incumbent on you to come up with a set of options to present to your stakeholders. You also have to realize that they will be coming at this in a highly emotional state. You're late! The project is over budget! And no matter what happened to get the project to this state, they will blame you and those obnoxious nerds you work with. It's not necessarily fair, but that's the way it will go down.

When it comes to options, my number one preference is to reduce functionality. Stay with the existing budget, stay with the existing delivery schedule, but reduce what gets deployed. Your client gets a tool, albeit reduced, that they can start using to improve their workday. No one has to go up the management chain of command and explain that they are in trouble, that they can't manage a project, that they are a failure. It keeps the engineers moving in the same direction without veering wildly off course. It forces an engagement with the users to prioritize functionality. They are now even more a part of the solution; they are engaged and active and have ownership.

Adding money, although sometimes necessary, makes all of the business owners look bad. The perception is they either willfully or through incompetence underestimated this project, overinflated their ROI, and are embarrassing themselves.

Adding time is almost never possible without adding money, but in those cases when you are behind owing to a lack of staff, it's the next best choice to reducing functionality. Although the ROI will not be realized as quickly, it's much less embarrassing than it would be to add money.

One other point about functionality is that it changes. In my experience, 60% of the changes that will happen with a system occur within the first year after deployment. This is largely because users see what the system can do and add features to take advantage of the new efficiencies. Conversely, they also see that the system does what they asked of it but not what they need it to do. It's perhaps not intuitive how a project can go through all the analysis, design, prototyping, development, testing, and deployment and still leave the users with something other than what they need, but it happens frequently. It can be difficult for someone buried in the current way of doing business to imagine operating with the new system, and they can firmly believe the system should behave in a certain way, only to find out once they use it that the new business concept has a flaw. Building out every function of a system before the users get a chance to use it, at their desks in a real world environment, is a very expensive risk. I'm a huge supporter of delivering a bare bones system and asking users to use it for a time and incrementally add to it as real-world experience informs the decisions.

Running a Status Meeting

The status meeting is possibly the most important interaction you can have with your team. It is when you check in on the adherence to the approved priorities, when you learn where the team is with their tasks, and when you hear about problems. Well, that's the theory anyway. Most status meetings are resented by the team, dreaded by the manager, and don't accomplish a thing other than irritating everyone.

Timing and frequency of status meetings are important. I'm a proponent of the very quick daily standup. Some folks prefer a weekly meeting, but I argue that if you're only getting the detailed status once a week, you don't have time to correct anything before small problems become large problems. A weekly meeting on Friday afternoon is only useful for ruining your mood all weekend. If you insist on a weekly meeting, hold it on Tuesday, Wednesday, or Thursday. That way you don't have Monday holidays interrupting things, and you have some time to accomplish something during the week, get in trouble, and recover before the weekend.

Another issue I have with the weekly meeting is that it is resented every time and is a dramatic break from the work routine. I prefer to hold a 15 minute (or less) status meeting about an hour after everyone has been at their desk and had some coffee. For your typical development team, that's around 10:30 or 11:00 AM. The average developer is "red shifted." They generally come in later in the morning and leave later in the evening. You have to adjust it to your specific crew. I've worked in some environments in which the engineers didn't arrive until nearly noon and others in which showing up at 6 AM meant that there was no coffee left for you--but the latter was unusual.

The goal of these status meetings is to set up a safe environment in which everyone can very quickly report accomplishments, lay out plans for the day, remind people of outages, and in turn be reminded of items that might not be written down anywhere. For example, this is a great status report from an engineer:

> Manager: "Bob, how's your stuff coming?"
>
> Bob: "Yesterday I finished the connection to the database. I'm stubbing out the reporting logic for now because we're waiting on an answer from the users on the final report. Once I get that I need about a half day to finish it. I have a dentist appointment at one, should be back by three."
>
> Manager: "Are the users aware that they owe you an answer?"
>
> Bob: "I think so. They should. I mean, I sent them an email last week."
>
> Manager: "Okay, let me take the action to get that answer for you. This is due by the end of tomorrow. If I get you that answer by the time you're back from the dentist, will that work?"
>
> Bob: "Well, yeah, I suppose. As long as Peter's stuff is ready."
>
> Manager: "Peter, you all good with that?"
>
> Peter: "Mostly."
>
> Manager: "Okay, we'll come back to that in a sec when it's your turn. Bob, anything else?"
>
> Bob: "No."

This was a great exchange for the manager. It was quick, identified two dependencies and one resource outage, and resulted in actions getting assigned and deadlines for them set. Okay, you have to take my word for Peter's stuff and associated actions, but you see how it's quick, safe, and there are no sidebar discussions. No dumping on the users who've had a whole week to get them the answer but haven't bothered to worry about it yet.

As the manager, you have to keep it focused on the work, not the personalities. You have to identify the problems and keep it safe to do that. Frankly, I think Bob is not going to make tomorrow's deadline. He's never going to get back by 3 PM from a dental appointment that starts at 1. He'll be lucky to be out of the chair by then. Dentists are notoriously late, with the exception of my current dentist who is always on time and would never harm me for pointing out this failing in his cohort. The fact that Bob needs someone else's unfinished code means that he hasn't yet tried to work with it, discovered the disconnects, and corrected them. Finally, Bob is working on a report, and reports take an insane amount of time to finalize. Users are always fussing with them up to the last minute, so Bob is going to be late and the manager now knows this and can decide what to do about it.

If the team were in a good rhythm, Bob would have mentioned that user dependency last week when it first came up. As the manager, you should have been following up on that. Don't ask your developers to be the ones to hound the users for answers; find someone who is more diplomatic. If you have a business analyst on the team, that person is often the best choice. All too often I've seen the engineer's curt and factual email anger the user community and set up a defensive, finger-pointing relationship such that if things ever go wrong the two groups blame each other. Users will resent the way they were spoken to during the project and that they have no give when it comes to last-minute functionality reductions that might be needed to get the release out.

One last point: Peter said his work was "mostly" done. That means "not" done. "Mostly" is engineer-speak for "No, Not, Uh Uh." Peter is also in trouble, but he hasn't spelled out yet exactly how deep in the mud he is. From a meeting perspective, you don't want to be pinballing among everyone in the room. You need to find the problems, make sure everyone's on task, and then let them get back to their tasks. Notice I didn't immediately say, "*What!* Peter, are you late!?" Instead I said, "Okay, we'll come back to that in a sec." Wait until Peter's turn to report. Keep the structure of the meeting, which is a single orbit around the team with each person reporting their status, after which they can leave. If problems need to be discussed and solutions found, that is not done in the status meeting. That is not status information; it's engineering and design work. Ask the interested parties to get together on their own and solve it. Keep the meeting short and sweet and predictable and safe.

Safe is key. It has to be safe for an engineer to say, "I'm going to be late because…" "I'm having a hard time with…" "I estimated this too low because…" "This is turning out to be harder than I guessed…" If it's not safe, then very bad things can happen.

I was once called in to help bail out a project that was in serious trouble. Trouble with a capital T. The country was in a recession, and jobs were hard to come by, particularly in that area. It was this work or painting lines on roads somewhere, so hanging onto this job was the only chance for professional employment. It was a brand new team, hired for the specific project, and with any luck they'd be together for five more years. It was a brand new management team as well. They'd not worked for this customer before, nor had they worked with the engineers. Everyone was learning.

There was one engineer who slid in, barely, on the interviews. He was not considered the best, but he was thought to be good enough and was willing to work cheaply, so he got hired. Unfortunately, he was not up to the task. He knew it, and he was desperately trying to lay low and learn on the sly both what he needed to know and what he had said in the interview that he already knew, while somehow making his deliveries in the meantime. If only he could get through this next delivery—if only! He truly did not know how to do the tasks assigned to him, but he did not feel it was safe to say so or to ask for help. He figured if he did, he would be fired, which would mean the end of his career in software.

No one had clue that he was struggling. Instead he did what he knew how to do. He was in charge of the user front-end, and sadly, all he really knew how to do was to place characters on the screen, look for user changes, and post those on the screen. What he built was a complete mock-up, a fake, a sham, with no operational code behind it. It was about as functional as the clay "concept" car mockups you sometimes see at auto shows. His inadequacies weren't discovered until the team was in the integration phase of the project, and by then they were already three months late for completely different reasons. This was a three-alarm fire. The best engineers from all over the country were called in to bail out this team and write code in a couple of weeks that should have taken months, plus senior management came out to smooth the feathers of the customer. The company only barely hung on to the contract.

The point of the story is that if it's not safe for someone to ask for help or relay a true status, nothing good can happen. You can control your own language and your own temperament in the status meeting, but you can't control everyone else. We've all seen meetings go ugly when one of the engineers throws a barb at another one. Yelling, screaming, name calling, hair pulling, and Klingon death curses will follow. So you have to also set the tone for the team. "This meeting is to make sure we all know what each of us is doing and how it affects everyone. It is where we identify problems and set a time to go about solving them. It is where we help each other. None of us is perfect. All of us will have a problem at some point in the project, and all of us are responsible for helping all of us. We will succeed or fail by how well we support one another, not by how well we individually do our assigned tasks. That's the easy part. Doing your tasks *and* helping someone else is the hard part." So now you've laid down the standard. Doing your own tasks is expected and assumed. Of course you're smart enough to do that. *But,* are you smart enough to do your own tasks while helping someone else? Raise the bar; make sure they know that support for someone else is rewarded, recognized, and considered the standard for brilliant. More help will yield them more prestige.

You also have to cut the sniping immediately, publicly, yet still gently. "Bob, that's not helping. I know you're waiting for this service to be done to complete your own work, but sniping at Peter won't get us there." Do not immediately ask Bob what he'd do to solve the problem. You're likely to get some snippy answer like "Give it to a real engineer" or something equally productive. Besides, Peter won't listen to Bob right now. He's mad at Bob. Bob might have a solution straight from the Creator, but I promise you Peter will not implement it, because it came from Bob. He'll spend all weekend making some half-baked solution sort of work before he'll do what Bob suggests. There's more on settling disputes between team members in the chapter "Dealing with Fights on the Team." My point here is that you can't let fights start in the status meeting, or we violate the goal of making it safe. As an engineer, if I know my intelligence can be attacked during the status meeting, I will never reveal that I'm struggling with a task.

Another standard you have to maintain—and this one is tough—is that the status meeting is for identifying problems and assigning people to solve them, but not to solve them. This is truly difficult for engineers. Problems present opportunities for their minds to light up like starburst fireworks. Everyone wants in on the action. I don't have any scientific research to back this up, but a problem seems to trigger the pleasure centers of an engineer's mind—it's like chocolate. The engineers will almost immediately lose all focus on the status meeting and focus on the problem. That's great and wonderful, but it defeats the goal of the status meeting. You have to quickly, very quickly, step in with the dreaded, "Okay, good, I think we know what the problem is. Bob and Peter, will you guys work on that right after this meeting? Anyone else who wants to contribute, please do. Moving on…" You will see shoulders slump, pouting, arms crossing across their chests, and other body language signs of distress and disappointment that delayed gratification can inspire, but you need to do this. Most problems will take at least fifteen minutes to solve and you still don't know how many problems will be identified today. You also need to get the team back at their desks and working. In a lot of cases, half the team's day might be wasted by trying to work a solution in the meeting when some members have nothing to contribute and should get back to work. So you have to be the mean parent who won't let the kids play just yet.

Surprises in a status meeting are bad. You need to emphasize that surprises are bad. All surprises are bad. Good surprises are bad, too. This was taught to me by my dear friend Stan Lucas, and he is absolutely correct. At the beginning of the project you need to get out your soap box and start preaching this gospel of projects. "If you are going to need more time to accomplish some task, I need to know as soon as you realize it. I can adjust things if you tell me ahead of time. If you come to me on Tuesday afternoon and tell me you won't make a Wednesday morning code freeze, there's nothing I can do to help you. We're sunk. Tell me the previous Friday and I can perhaps get the user functionality simplified, maybe we can work together to figure out a quicker, albeit not as robust, solution, or maybe we can get more time. But the key is you need to tell me so the team has some time to work out a solution."

Notice I didn't use the word "late." Late to an engineer is a failure. They weren't brilliant enough to deliver on the original schedule. We hope it was the engineer's own time estimate, but even if it wasn't, the engineer will see being unable to accomplish the assignment in the allotted time as a personal failure. Someone thought this was enough time, and now I'm not up to it.

Notice I said "…maybe we can work together to figure out a quicker, albeit not as robust, solution." This is targeted at your perfectionists, your constant code refactorers, or someone who won't let go of the original foundation even when the business rules change. I didn't say, "Maybe we can get your code to work by getting some smart eyes on it." I didn't say, "You're making this too complicated; it's a simple problem, what on earth are you doing?" I didn't say, "Stop polishing this apple; you've got it down to the core!"

Finally, notice that I didn't say, "Add more time" until the very end. That's your path of last resort. The schedule and estimates need to be treated as real and fixed. The schedule isn't a pirate's code "guideline." It's very very real (yes, two verys!). If you let the schedule slide this way and that, there's no reason to worry too much about the estimating. If the estimates are given short shrift, you suddenly have no idea how your project is going and you are now set up for a hellish ride to a colossal failure.

"Finishing early" surprises are not usually as bad as finishing late, but they also create problems. Finishing early should be a good thing. You're ahead! But you're only ahead if you can stay ahead. You need to know early if someone is going to require less time. Sadly, this is not a frequent problem, but it does happen with a couple of tasks on any given project. The nature of the problems it generates is not the same. You are now trying to capture and preserve that jump on the schedule rather than just lose it in the sauce. The only way you can retain that schedule advantage is if you can bump up all the other dependencies in the schedule along with it. This is often possible, but only if you have advance warning. "Guys, if you see that you can finish a task early, let me know. The only way we can retain that jump in the schedule is if I can bump up the testing, the user reviews, whatever, so I need time to arrange it. It's a wonderful thing either way, but it can be our saving grace if we can hang on to it. It will bail us out when we lose the servers for a day when they do some unannounced patching or the network is out."

Notice I did not say, "It will help out when you're late on some other task." I pointed to potential delays not in their control. You don't want to ever imply that they are not going to be able to meet their schedule, which is insulting. As soon as you do, they've turned off to listening to you.

Ending the meeting and moving on seems simple but actually requires some thought and, frankly, some ritual. You need to make clear that this meeting is over and that it's time to get to the action items and normal work. Many engineers don't like to take notes in meetings. Keeping things in their heads is a core competency of developers, and they don't often need to write details down. So I like to end the meeting with a recitation of the day's action items and some little catch phrase, much as Walter Cronkite used to end his newscast. "And that's the way it is. Goodnight."

> "Okay, good, Bob and Peter, you two are going to work on the database problem. Peter, you're going to work with Bob on how to handle the potential errors the code could get back from the database, and I'm going to track down the answers from the users. Thank you everyone. Let me know if anything comes up today."

And then I stand up (if seated) and walk out of the room. You need to initiate the physical end of the meeting.

By the way, I talked about the status meeting being a standup meeting but I actually like to sit down. Unlike good engineers, I have a terrible memory. I need to take notes and write things down, and I'm not good at writing on a pad I'm holding in the air.

Baa Ram Ewe: How to Get an Engineer to Take an Undesirable Task

In the movie *Babe* the old ewe named Ma tells our hero, the Pig named Babe, how to communicate with strange flocks of sheep. She tells him the pass phrase, a poem, actually, that apparently all sheep use. It begins "Baa Ram Ewe." There's a similar phrase that will force engineers to take on a task, even if they don't want to. It's a mesmerizing and hypnotic phrase that will only work a couple of times, so you don't want to use it unless you're truly up against it.

Once in a while there's a task that you need someone to do, but it's not glamorous or it's tedious or no one will ever see it and therefore there is no glory in it. Data conversion often falls into this category. Building the test database is another. It's the kind of task you know will get no volunteers. Instead of asking for volunteers, ask them instead, "Do you guys know anyone smart enough to handle this task? It's very important and we have to have it done just right."

Frankly, it doesn't matter whether they know you're playing them. It's like flies to honey, moths to the flame, opposite poles on a magnet, chocolate and peanut butter—you name it. The gauntlet is down and engineers just cannot let it lay there. They might be able to laugh it off at first, but in a little while they'll be forced by their DNA to come back to you and tell you they're up to the task. Why? Because they *are* smart enough to do the task, and leaving the task undone would be like admitting that they are *not* smart enough. No one wants to be thought of as less than able to handle the task, so they will be forced to pony up. Use it when you must. Really need to when you do. This type of powerful magic should not be used lightly.

Explaining Changes to Engineers

As I mentioned earlier, engineers are most often judgers. They like a plan and its associated goals. It's a nice, neat, defined problem to which they can apply their talents. However, things can change. Management can reprioritize things. Users can change their minds. There are many reasons why things change once the project is under way. The issue for a manager is that engineers hate changes. Again, their goal is optimization, efficiency, and cleverness. How can one possibly achieve any of this if the fundamental goals of the effort change? Much of their previous work is now useless. Not only have you defeated their goals, you've wasted their time. You did not appreciate what they did for you. You told them that the x function was the most important thing to the organization. They went ahead and started to build you the best system around to optimize that. They poured their souls into it. Used all their talents. And now…you are negating all of that. They get no acknowledgment of their brilliance. No pat on the back…nothing. They can get angry, depressed, de-motivated. From their perspective you asked for a ranch house. They constructed a perfect foundation and now you want a bank with a drive-through window. Why can't the users make up their minds?

As the manager, you have to take the team through a process that is a lot like grieving. You have to explain the situation. The client has put the project on the shelf, and we now have to focus on this new project. We have to stop working on the mobile module and instead focus on the reporting module.

The reaction will be denial and disbelief: "What are you talking about?" "This makes no sense!"

Next will come the anger: "This sucks!" "Are you kidding me?" "They pulled this crap on us last time!"

It's okay to let them vent for a time. They are hurting. All that effort, all those great ideas, the patient attention to a thousand details—all of it is gone, wasted. Delayed at best. At the same time, you have to cut off the venting before it becomes toxic.

Next comes bargaining: "They should just let us finish this. Do they realize what this will cost them? We can't pick this back up and move ahead and keep the same pace later on!" "Can they extend the deadline and let us do both?" "Why don't they get those other guys to do the reporting module?"

This is real for them. You must respect their passion and their loss. On very rare occasions there are options but usually not many. So we have to move on to acceptance.

Acceptance happens when you see them refocus on the new challenge. When they start asking about the new features needed, the business to be supported—that's when you know the team is moving forward again.

There's a temptation to blame the users for changing their minds. There's some truth to this; they did indeed change their minds! But they likely didn't do it maliciously. They didn't set the team up for a major change and then go off to some coffee shop and laugh about it. They are learning more about their own work and making changes based on that. They are learning what could be done if the system were automated, and they are responding to those new or future capabilities. They are also human and can get themselves stuck in group think and go down paths that they later realize aren't useful.

The user community is often being pushed or dominated by a single alpha personality. If that alpha is reassigned or is out of town for some period of time, the Will of the Masses might re-emerge and the changes flow down to the engineering team. My point is that it's natural, happens on every project, and makes the engineer's life hellish.

There is a temptation to say things like, "Well, they just figured out what they need to support a business they've been doing for 30 years!" or "Hallelujah, they raised enough money to buy a clue," or other equally dismissive comments. You've now sided with the engineers, and they might stop yelling at you, but they will still be mad at losing all the progress they made in the original direction. Taking sides will also do nothing to bridge the divide between the business side and engineering—instead it probably burns down those bridges.

Instead you need to help the engineers understand the problems from the business side. "Guys, the users are just now learning what you can do for them. What the tools you're giving them could mean to their business. They've never had this capability before, and they're working their way though what it means to their processes. We sure don't want them doing things exactly the same with the brand new system. That would be a waste of time for everyone. So yeah, they're changing things to try to use these new tools more efficiently."

Or "Guys, the company took a huge budget hit. Their sales are down, they had to cut the budget, and they had to decide what the biggest pain point was for them. In their minds, the biggest pain point was reporting. They thought they needed more insight into what's happening day to day, even if it meant that they had to do the data entry the old manual way. They decided that the potential savings our new automated entry would provide were exciting but not as strategically important as the reporting. They felt that once they had a better handle on the inner workings of their business, they could move out with the specific data entry problem."

The team will want to know that you presented their arguments to the business side, as you should have as a responsible project manager. "I told them that it would take more time to pick this up later, and they said they understood and would accept that cost." "I told them that this will require changing everything we've done so far, and they said they understood but it had to be done." "I told them that this would limit what they could do down the road, and they said they'd deal with that later."

If you keep in mind that you are asking them to accept abandoning their child on a rock with the promise that you might come back to get it later, you'll be able to get through their emotion and refocus on the new assignment.

One final approach if you're truly stuck with the team fighting things. You can explain, as my dear friend Stan Lucas taught me, "The customer isn't always right, but they always get to make the call on what we do for them."

XX Chromosome Engineers: The Forbidden Chapter

My wife told me not to write this chapter. She read it and told me to take it out. I'm sure she's right and I'll regret this, but I just felt that ignoring a discussion of some similarities and differences I've noticed working with women engineers would leave the book incomplete.

I'd like to be writing this based on an absolute wealth of knowledge of and experience with the legendary female engineer but (1) I'm a man, so I can't ever say I have knowledge and understanding of the opposite gender; and (2) even here in the twenty-first century there is a dearth of female engineers. I'm not convinced that the numbers are all that much better now than they were in the early 1980s, although nationwide statistics indicate otherwise. Female engineers not as rare as unicorns or Bigfoot or even black swans, but it is hard to walk into an IT organization and find it littered with women. So instead of speaking boldly and covering a topic thoroughly, I'll share some experiences I've had and a few potential insights that need more data before they become teachable moments.

Female Engineers Are Good

Guys will hate this, but my anecdotal evidence suggests that the average female engineer is better than the average male engineer. I'm not saying that all female engineers are better than all male engineers. What I'm saying is that I've found very few female engineers who were just average and none that was bad. Honestly none. My nonscientific theory on this, based on nearly no research, is that software engineering is still a male-dominated world. If a woman wants to enter this world, the deck is stacked against her. The result is that women have to really want it, and they have to work their tails off to get the same recognition and the same level of achievement as their male counterparts. To be accepted by the guys, they have to be at least as good as them and usually better. Drive seems to be an excellent teaching tool.

Female Engineers Don't Often Have the Same Hardware Experience

When parents see that a son is mechanically inclined or interested in how things work, they will usually give him a building set (e.g., Knex, Erector) or one of those electronics science kits with which he can make radios, solar powered lights, three-way switches, and the like. Boys are given more hands-on practical knowledge of physical things. Almost every male engineer I know had one of those electronics kits growing up. My dad and I built a bunch of Heathkit radios, which included winding coils for the transformers and tuners. We soldered up the printed circuit cards and assembled the bits and pieces into working radios. By the time I was ten I recognized all the major types of electronic components and what they were used for. I knew about the ratios of the coils in a transformer, how electric motors worked, and how to use a soldering iron. My older sister was given none of these experiences. My wife, a senior software engineer and a fellow physics major, first saw a resistor in her sophomore physics lab.

The result is that they can be behind their male peers in their knowledge of, and often uncomfortable around the mechanics of, computers. Male engineers love to rip open the covers of their PCs and swap hard drives, put in memory, anything that lets them play with the hardware. Female engineers don't always have the hardware knowledge, and they lose face, status, and confidence when they have to turn this problem over to their male counterparts. Sometimes they will open up the case only to hesitate when not sure of what they are looking at, and stop. Guys love to help out and demonstrate their mastery of the machine, braving the danger of the coursing electricity and data! They usually don't explain what they are doing or make it a teaching moment as they would likely do with a male engineer. Instead, with the best of intentions, they do the "don't you worry your pretty head there, little lady" style of rescue. So the women still don't know how to do it. It's hard for the women to learn from these moments, too. Asking detailed questions about the hardware reveals one's ignorance and can result in a loss of status in the group.

Guys Grab Things, Women Wait Their Turn

Okay, another dangerous statement, I know. It has been my observation that both boys and men are perfectly okay with reaching out and grabbing/taking a new widget that is set on a table. I see this with boys and girls working on Lego robotics competitions when the boys just grab for the cool new Lego part or the partially built robot while the girls tend to wait their turn. I've seen it in the office when a new computing widget is brought into a room. Doesn't matter what it is—a USB hard drive of a kind that everyone has seen before or a brand new type of chip in its packaging—the guys will reach out and grab it long before their female counterparts will. Guys will grab it from each other, too. For guys this is normal, albeit annoying, behavior. In guy rules, it's okay to grab it back.

Guys will grab the widget from the women, too, but to women, this is a bigger deal. They react differently from the guys and are more likely to be insulted by this behavior.

In the Lego robotics groups, the girls often retreat and stop interacting with the new widget at all, thereby becoming less knowledgeable about how their group's robot works. In the office you see the women pushing back from the table and crossing their arms while the guys are busy jockeying for an angle to grab it back the first chance they get. Again, they lose the knowledge of the new widget and remain behind on the hardware side of things.

Woman Can Facilitate Team Communication

Men are blunt when it comes to their speech, more so when they are arguing. It doesn't matter what the argument is about—a sports team, a political position, or a potential solution to a problem—they will raise their voices and call each other names. Male engineers bring blunt to a whole new level.

Engineer A: "So to do this your way would be insanely stupid."

Engineer B: "No, what's insanely stupid is doing it the old way when we already know it doesn't work for high bandwidth environments."

Engineer A: "But we don't have a high bandwidth environment. This isn't that kind of system, which you should know by now!"

And it just keeps going round and round, and frankly it makes no difference whether a woman is present. Guys will do this whether there's a lady in the room or not. It's what we do.

Sometimes having a female voice in the crowd can calm things down. Here's another dangerous statement: women are better listeners. During the fray they seem to be able to hear both sides and still process the ideas, even with all the hyperbole. Also, with the guys going back and forth and getting louder and louder, they are often not participating in the screaming match. They are sitting there listening and thinking about it and...waiting for the guys to get winded enough to get a word in edgewise, which is hard. Guys are fine with interrupting each other. They step all over each other's talking. Women generally don't do this, except when they are saying something supporting a woman's statement, say, like finishing the sentence in unison.

Sometimes when there's a lull in the storm, you'll see the female engineer break in and say something like, "Wait, so what I hear Bob saying is that if we had a high bandwidth environment we should go with his approach, and I hear the other Bob saying we don't have that problem so we should do it his way. When I read through the documentation, I think it's unclear what bandwidth environment we have. It sounds like it would start out low and eventually get to high, but probably not for years. Is that right?" The calmer voice, the more soothing tone, can sometimes bring the heat down in the argument.

Women Can Be Hurt by the Way Men Debate Things

As I mentioned earlier, guys can call each other's ideas insanely stupid, and it's no big deal. They can call each other insanely stupid, and it's a bigger deal but usually quickly forgotten. Women don't generally operate this way. Women are a lot nicer to each other. If a woman disagrees with another woman, she will almost always preface her response with a "I understand what you're saying but..." or "That could be true but..." Not so much with guys. They just lay into you. If a woman were to act that way to another woman, it would be a serious breach of protocol. It would be considered mean and disrespectful. Woman still debate points hard and passionately, but there's more tact in the way it's done. They might attack the point but never the person.

They also make sure to point out a positive before they make any criticism lest they be perceived as mean or nasty. These rules are fine and dandy until you cross the streams.

I asked a neighbor of mine who had a lot of experience writing grant proposals to review one I wrote for a neighborhood project. I know technical proposal writing but have no experience with community project writing and even less with how to write a winning grant proposal. She, however, is an expert. She very graciously agreed to help out, and she made the proposal much better by doing so. The funny part (now) was our interaction on the review of my text. We went paragraph by paragraph, and she had to begin with a compliment or say something positive about each before she would point out that I had the wrong verb tense or that a sentence needed to be cleared up. She complimented paragraphs that had no issues at all.

> Sue: "Oh, I really liked the way the next paragraph lays out the ideas for the work. It was complete and very clear."
>
> Me: "Thanks, any problems with it?"
>
> Sue: "Oh no, it's great"
>
> Me [thinking to myself]: "Then why are we talking about it!?"
>
> Sue: "The next paragraph is also great. I liked how you described the community involvement and who would do what. I did notice one small thing, and you probably would have found it yourself, but the name of the association has an extra 't' in it."
>
> Me: "Got it. Good catch." [But I'm thinking "Holy schmoley, lady, just tell me there's a typo in 'communitty' and let's move it along!"]

When the worlds cross, it can be difficult. Sue was doing it right by her protocols. She was making sure that I knew she appreciated the effort I had made on this document and that I should be proud of it, as she was proud of me, and that I should not feel bad if there was an error here and there. We all make mistakes; it's only human. Meanwhile I'm thinking, "Why are we not going fact to fact to fact?" What's with all this touchy feely stuff? We each expect to interact according to protocols of our tribe.

In the office the blunt nature of guy conversation can be very hurtful to a woman expecting to be better supported. She can interpret this behavior as representing a lack of respect for her abilities—as personally insulting. It undermines her confidence.

Men Are More Assertive in Their Speech

As a general rule, when a bunch of guys are debating something, they will phrase their points of view as bold assertions. They will say, "This will be the fastest way to do it" or "This has the fewest possible hits on the database" or "You have to do it this way or it won't work." Now in most cases these assertions are provably wrong. You could pick each argument apart and find where it is flawed. That's not the point. The guy is thumping his chest and putting an idea out there. The rest of the guys know this. Women might intellectually know this, too, but in their tribe, people don't behave this way. People only make such statements when they absolutely know them to be true. More important, women would not phrase their point this way at all. It would be more along the lines of—and I'm exaggerating here to help make a point—"Well, I know there are a couple of ways we could approach this and there are possibly, maybe even probably, better ways, but I was thinking that one way might be to ..."

In the female tribe it's rude to pound one's chest and assert perfection. It's insulting to the audience. The guys hear all these caveats and they turn off. If an argument is so weak as to be only one of many possible ones, then it's not worth hearing. They dismiss it before they even hear it. The result is that the female engineer is shut down. She is now even less likely to offer up ideas if these jerks won't even hear her out.

Erosion of Confidence: Death by a Thousand Cuts

I've pointed out a couple of ways that gender behavior differences can affect how a female engineer feels about her work and herself on a project. Although in some cases the "rude" behavior was intentional, in most situations the slights are not intended. Guys can actually come to blows about something and then go out and have a beer. Most women do not respond to insults, yelling, and aggressive debate in the same way. They feel that they themselves have been attacked. Even though they can tell themselves that that's just their way, it still hurts. Each one of these interactions can take a small bit of wind out of their sails of confidence. Over time their confidence can be eroded and with it, their effectiveness in the office. Sometimes you see a brilliant, talented engineer who in her first few years was full of great ideas and suggestions become a follower behind someone you didn't think of as having the same potential. Sometimes women start to believe that the louder, brasher, bolder men are more capable or, more to the point, that they are less capable.

Sometimes Oil and Oil Can't Mix

My company won a contract to build a very complex system, and it was truly going to tax our technical abilities. I lobbied hard to get two engineers with whom I had worked in the past who were absolute stars. They both happened to be women. I was ecstatic. By having both of these stars on my team, I felt I had greatly increased my probability of bringing this project in on time and with a great product. I assigned each of them to lead a portion of the project. I actually slept better that night. That warm and wonderful feeling lasted less than two weeks.

Ten days into the project one of the two came to me and told me she had to leave the project because she could not work with the other lead. You could have knocked me over with a feather. Both of these women had led teams of people before. They had never failed. Now, suddenly there was an insurmountable problem? One was willing to leave the project rather than work with the other? Something must have happened, and I was determined to figure it out. Mind you we'd only been on this project for a little over a week, and honestly not a lot other than getting our environment established and on-boarding the team had happened, so I had no idea what the problems could be. She said that her counterpart "had to have it her way, that she wasn't letting her have any input and was rude." I asked the lead to hold off for a week and let me see if I could fix it.

First I talked to the other lead and asked her how things were going. She told me it was great but she was really having a tough time with her counterpart. I asked her what was going on and got back a "She's being difficult, unsupportive and insulting." I asked her for specifics, and she said it was mostly her attitude. "She's rude, won't listen to my ideas, she's pushy. It's not going well."

I held some meetings with just us three and tried to talk it out. I met with them again and again individually and never could get anything that looked like a smoking gun. These two women just hated each other and would not work it out. Eventually I had to let the one off the project and replace her with another engineer, who happened to be a guy, and we got on with the project.

I have a theory about what was happening here and has happened less spectacularly to me on a couple of other projects. It is only a theory. Women who get to be senior technical leads have spent years working with a largely male cohort. To remain confident enough in their own abilities despite the thousands of cuts they receive over this time, they have had to adopt their cohort's behaviors. They have learned to interrupt the guys in the meeting to let themselves be heard. They have learned to stop caveating their ideas in a group setting. They have learned to be blunt and to be loud. They will grab the widget back out of one of the guy's hands. They aren't one of the guys, but they can behave like a guy when they need to, to move a project along in the right direction.

But then they get plunked into a project with another woman, and they are not behaving the way women are supposed to behave with one another. They aren't "listening" or "supportive" or all the things that women often do in their speech that guys don't. It's now considered rude, insulting, not listening, pushy, and dismissive. If you're a woman and have gotten to a senior level managing guys, it's hard to flip the switch back and forth while still inside the office.

What About All-Women Teams?

I've managed a very few projects with only women engineers. As I said before, the odds are against this happening given the population, so my data points are few and far between. The standard deviation on the outcome is also enormous. Most went fabulously. Without the male influence, the women interacted as women are wont to do. They were free to use their female speech style and worked with each other cooperatively. When they had disagreements, they handled them tactfully. Our reviews took a little longer than they might have with a mixed gender team owing to the voicing of the positive before pointing out a flaw, but things went well. The teams were happy and supportive, and we finished the projects on time, within budget, and with happy customers.

One project went spectacularly badly. It was a small project and in an area that our crew had a lot of experience. It should have been a gimme. The team was young but very good, and they had been successful individually on a number of other projects. It was a small team: three engineers, a tester, and me. We should have been able to bang this one out with no problem, but it quickly devolved into open warfare. Not between the engineers and the tester, as you might have assumed. No, the battle was between two of the engineers ganging up on the third. Truly horrible behavior from the team. They would insult her. Shun her. Make fun of her clothing. Make fun of her boyfriend. Make fun of her code. Hide licenses so she could not upgrade her machine. They would come to me individually and tell me she was doing a bad job, that she was late with her stuff. That it didn't work. I'd check the repository to see when things had been checked in and couldn't find things that were late. I checked with the tester and her stuff worked. Our meetings were openly hostile. One day we went to a birthday lunch for an engineer on another project. When the picked-on engineer got on the elevator, the other two stomped off. It was childish to say the least.

One day the cabal of two came running into my office to explain to me how her stuff was not working and to demand to know what was I going to do about it. "You've been defending her all along, but you can't deny this one." I found out that the cabal of two had renamed a key library and moved it to another area so that when the odd-woman-out finished her stuff, it was linked to the old library and did not work. A complete, intentional sabotaging of the other woman's work. I was livid. I waited 45 minutes to calm down before I called the perpetrators into my office. I talked through what I had learned and how what they had done was not acceptable behavior and what behavior I did expect, and after about 15 minutes of me talking through this situation I had an unexplained pain in my right hand. I looked down and realized that I was pounding my desk with my fist and then became aware that I was shouting. Apparently I wasn't as calm as I thought. That was one of only three times I have ever yelled in the office. It had no effect, as yelling never really does.

We tolerated each other through the finish and did turn out a good product in the end.

What did I learn from this experience? It was such a fluke and such an outlier that I'm not sure there's a real message here. I suppose sometimes women can be nasty to each other. It's never happened since, so I have nothing to compare it with. I hope it never happens to you.

So What Should You Do?

I've talked through a set of experiences and differences between men and women in engineering. How as a manager should you deal with them? Each engineer is a resource to be treasured, and it's how you nurture that treasure that determines the quality of your team in the long term. Honestly, the things I recommend here are the same things I'd recommend for a quieter, shyer male engineer whose input you want and contributions you need.

- **Make sure that everyone is heard**. This means that sometimes you have to stop a debate and turn to that silent person and say, "Betty, you've been listening to us argue, what do you think?" Now the floor is hers, but you will have to cut off the interrupters who are probably very used to cutting this person off.

- **Set up ground rules for meetings**. This sounds obvious, but it's good to repeat them at the beginning of every design session or potentially contentious meeting. Those ground rules need to include the following:

  - You may critique the idea but never the person.

  - You may make no personal remarks or slights.

  - *No interrupting*. Now this last rule hardly ever gets followed, but it's a good goal.

- **Don't let people grab things**. This is simply a matter of playing referee. Stop people from grabbing things just as you would with a group of kids. "WAIT, Bob wasn't done with that. Bob will give it to you when he's done." In fact always give the object that is likely to be grabbed to the least aggressive person first. Might be a woman, might be that quiet guy. Doesn't matter. When you do, make a point of saying something like "Bob, take a look at this and pass it on once you're done looking at it."

- **You ask the hardware questions**. No one expects the
project manager to know the insides of a computer, switch, or
disk array. Heck, they're not sure you can tie your own shoes.
If you're standing with someone you think won't ask questions
they are dying to ask, you ask. Even if you do know, ask. Make
someone describe what the various components are, what the
functions are, how things work.

If you don't take pains to referee the interactions on your team and
help those quiet folks, you run the risk of having talented engineers
doubt their abilities to solve hard technical problems. Once they start
doubting themselves, they either become follower engineers or drift
off to management, testing, or requirements analysis and other more
people-oriented things. They don't contribute their technical talents to
the team, and they give up on their dreams and goals of succeeding in
the technology field.

Testing

By the time a piece of code enters the formal testing phase, the project
is often in complete turmoil. Testing is usually the last step before
deployment, and since it is there, over on the right hand side of a
Gantt chart, it is often compressed and abbreviated. All the other
delays in all the previous phases of the project have now come home
to roost on the shoulders of the testing phase, so it is often a rushed,
cursory checking of the system to demonstrate that it will work. This
is exactly what testing should never be.

The purpose of testing is to find out where the system is broken so
that the project stakeholders can make an informed decision about
what they want to do about it. Is it a show-stopper that must be fixed
before deployment? Is it something that won't be used for months so
it can be fixed in the interim, or is it something that we can just live
with? No system of any consequence is unbreakable. It is broken
somewhere, and testing is supposed to tell you where and how easily it
can be repaired.

I assert that testing is possibly the most schizophrenic of the phases
when it comes to process maturity. Testing against requirements and
regression testing are very solid. There are excellent automated tools
that support testing and re-testing applications to ensure that the

desired outputs happen when sets of inputs are provided to the system. A good test engineer is a very organized person who will set up each of the scenarios for you and create a raft of test data. In the end you will be able to know for sure that the basic functions work as intended, which is perhaps 60% of the battle.

The "Sybil" side of testing is figuring out how to break the system. Here is where the process is immature at best. It takes an evil genius to figure out how to break a system. If I write a piece of code to check for numbers in what's supposed to be a Yes or No entry, that's trivial to develop and to test. It's the evil genius tester who will figure out a way to lock up a dictionary record just long enough for your update of a client record to time out but only after the order has been posted. That takes genius. That takes a wonderfully warped mind. It's this mind that will figure out how to find the places where records could be skipped during the nightly backups. It's this mind that will figure out a sequence of steps that no one ever considered that totally bollixes up the data. And more important, it can be done repeatedly.

A great tester is the analog to Donald Sutherland's arsonist character in the movie "Back Draft." He loved fire and looked into the eyes of "the beast." Your great testers are people who know programming, networking, database systems, system configurations, and want to see them burn. They are evil and your best friends. They will make your team look great, if you let them. Too often managers and developers see only the embarrassment of errors in their code or scenarios they never contemplated. They don't see how the tester has prevented the error from being seen by a broader audience or how they have now improved their game by showing whole new ways that systems can fail. Next time, their code will be so much stronger. They have improved their Kung Fu, not dishonored the family.

So, how do you help the tester help you? You need to help them be as efficient as possible with the standard stuff so they can turn their evil genius loose on the unstructured stuff. I have seen people turn whole systems over to the test team at the eleventh hour, and it was the first time the test team had ever interacted with the project. This last-minute strategy had the expected poor results. Instead of waiting, you should engage with the test team early and throughout. Your testers need to learn about the requirements of the project as early as possible. They need to design test scenarios and dream up the right sorts of test data. They have to monitor the size of the effort as much as any other function on the project.

Management often fails to factor in additional test team time when they make changes to the scope of the project. Testing will still be those last three weeks no matter what happens along the way. It doesn't make any sense, but it's typical for a project to lock in testing to a fixed block of time and never look at it again. The test team needs to see early drafts of the user stories or requirements documents. They should see the concept of operations for the system. They need to learn about the business that is supported by the system and the data they use in executing that business. They have to know the business cycles and the end-of-year processing that might happen. If you are so fortunate to have a tester who is also very good with the users and you also have the bandwidth, using your test engineer as a business analyst is wonderful. It gives the engineer the first-hand understanding of the business that their evil-genius persona can use to find ways to make the system fail. Unfortunately, many great testers were once software engineers, so they sometimes come with all the interpersonal skills that the engineers have (or lack).

Your testers also have to know about the architecture of the system you are building. They need to see how data flow through the system, what the controlling processes are, how they communicate, how they secure things, and the like so they can work on messing with these as well. Thus, the testers should be included in any design reviews. You also need to set up time for them to meet with the engineers and get a detailed walk through of the system.

The testers should see the user interface as early as possible. This bounds what types of data they can inject in their tests. It also provides a whole separate area for evil genius to work on—breaking the interface itself.

The testers will need a lot of test data and a way to refresh and restore the data. This task will almost always fall to your development team, so you need to allocate some time for them to support the test team in this way. If you were paying attention, you probably noticed that the testers would need a lot of time from the engineers to describe the system workings and structure as well. Scrimp on this time at your peril.

Testing is also possibly the biggest blue-sky guess of all of your estimates in the schedule. If you're building something new, how can you be sure of the time it will take to test, remove defects, and retest it? How long will it take to figure out the source of the defect and repair it? How many re-tests will be necessary? There are no reliable industry-wide metrics for this. The test phase is an intensively interactive time between the testers and the developers. There's an assumption that the developers will be removing defects at the same time that testers are moving on to other parts of the system. It doesn't always go that way. Oftentimes a tough defect to figure out will suck up the time of both teams working together. Also some defects might make further testing impossible, so your testing comes to a screeching halt until that one defect is corrected.

I've seen some people build project plans that assume the developer will be setting up the production environment while the testers are testing. That never works. The development team needs to be committed to testing for the entire duration of the test phase. It's just how it works in the real world. Finally, just to keep project managers up at night...you cannot know where you stand with testing until it's done. If you are ten days into a twenty-day test period, you have no idea if you are truly halfway there. Anyone who tells you different is lying.

Testing Can't Be Perfect

Testers can only test things that we can predict might fail. We build test cases for those and feel good that we've tested the system. There is a myth that if we just had enough time to test the system, there would be no defects in it, but this is false because testing is done with a subset of data. It is done in a separate environment from the operational system with a handful of users at best. There are tools to perform load testing, but again, that's usually done only on the test environment. Most organizations would not risk an outage by running it on their live system. So how can you test the live system with live users on real loads in the real environment without being on it?

The answer is you can't. Testing, like estimating, is your best effort.

You can do a crummy job, for sure, but you can never state that you have found all the defects in the system, particularly if you have a system that is sharing an infrastructure with other systems. Rarely does testing include putting load on the "other" systems while testing your own. The resources are so rarely there to make that happen. Even then you're testing a notional load with out-of-date data or a subset of data. Testing in your anemic test environment might show bottlenecks in one spot that disappear in the full production system, which has twice the processing power and completely fiberoptic network. Likewise, it might never reveal bottlenecks that happen when you have the full 5,000 production users pounding away.

In 1999 the National Aeronautics and Space Administration lost the Mars Climate Observer satellite when it was piloted too close to the Martian atmosphere and disintegrated. The cause was later called the "Metric Mixup."[123]

The satellite was programmed to use metric units, but the ground station controlling it used imperial units. Oops. How could this have happened? It wasn't an omission; it was an actual mistake. The contract for the ground station specified metric units. How could this not have come to anyone's attention? Two reasons:

[1] Arthur G. Stephenson, Mars Climate Orbiter Mishap Investigation Board Phase 1 Report, http://sunnyday.mit.edu/accidents/MCO_report.pdf (November 10, 1999)

[22] Wired News Report, Mars Mission's Metrics Mixup, http://www.wired.com/science/discoveries/news/1999/09/31631 (September 30, 1999)

[3] Virginia Commonwealth University, Metrics Mixup May have Doomed Orbiter, http://www.has.vcu.edu/new-phy/doomed.htm (1999)

(1) If you're testing the ground station, all of the calculations are consistent, and when hand checked they came out accurate. The testers were focusing on the accuracy of the calculations, and they were correct. The error was probably made very early when the first set of calculations was done. The space flight engineer probably was used to imperial units, and the testers made sure to check with him, and everyone signed off on it. If your entire environment is in imperial units and everything works, then fine. Asking if they were in the correct units never crossed anyone's mind.

(2) You can't test drive a satellite in anything like a realistic scenario. There is no environment to test launch and steer a satellite around until it's actually on its way to Mars. NASA has solved the first problem by requiring a set of tests to ensure unit consistency. It is now a predictable point of failure for their systems, and they have it covered. Before this disaster, no one would have thought of it. As a side note, two engineers did point out the discrepancy between the predicted and the actual trajectory of the Mars Climate Observer before it was lost, but their warnings were ignored.[4]

As the project manager, you want your test team to do their best to ensure that everything that can be tested in the test environment is done, but you also have to back them up when a defect is discovered in the production system. Senior managers are wont to take shots at the crappy job you guys did in testing. That's unfair and is based on senior managers' lack of knowledge of how things work. As the project manager, you have to be visible in making clear that this problem could not have been found in the test environment, but we'll get right on it now that we know about it. Never throw your testers under the bus, even if you believe that they will never hear about it. Senior management needs to be educated on what can and cannot be expected of the process. If you decide to try to save face by blaming your testers, then what happens the next time? About the third time you blame the test team, management will be expecting some heads to roll, and it's not the testers' fault.

---

[4] Wikipedia, Mars Climate Orbiter,
http://en.wikipedia.org/wiki/Mars_Climate_Orbiter (April 2013)

Software Engineers and Testers

By the very nature of the work, testers and software engineers are in constant conflict. If they were in couple's therapy, the conversation might go something like this:

Software engineer: "I don't even know why we're here."

Tester: "You and I are here because we have problems, *big* problems with our relationship. You never talk to me anymore. You just sit at your desk and code. There's no sharing. I have no idea what's going on with you."

Software engineer: "What's going on is that I'm busy and I'm tired and I have a lot to do."

Tester: "You came in late and with new requirements all over your code! You think I didn't notice? I didn't know where you were or what you were doing or who you were with, and you never called to tell me anything."

Software engineer: "I was working, the users changed their minds, and I had to stay late. I can't stop everything I'm doing to call you every time something changes. I have work to do!"

Tester: "Well, it seems your work always comes before my work. I don't think you appreciate what it takes to keep this project going."

Software engineer: "I don't? I don't! Really? I'm the one who has to make the code. I'm the one dealing with all those people and their crazy stuff day in and day out. I'm not the one sitting back here all nice and quiet preparing scripts to run."

Tester: "See! That's what I'm talking about! You have no appreciation for what I do!"

Software engineer: "Do? All you ever do is complain! Nothing I ever do is good enough for you! You pick and you pick and you pick. This isn't right or that isn't right. There's no pleasing you!"

Tester: "No pleasing ME! I'm working all day long trying to make a nice release for our project, and you go on inviting changes at the last minute and I'm supposed to put on a full test suite and be cheerful about it! And I do it! I do it because I care about our project and our team. And our team is coming apart!"

Software engineer: "If it's coming apart, it's because you're pulling it apart. I can't be doing everything around here."

Tester: "Everything!? My work is important to our project, too. You wouldn't be able to deliver unless I did all that I do to make this system nice. Oh, and by the way, would it be too much to ask to do your own unit testing? Is that too much to ask? Could you pick up your own code now and then?"

Software engineer: "You're just going to redo it. You don't trust me to do it myself anyway!"

Tester: "I trust you but don't forget about that other system."

Software engineer: "Oh golly! Are we on that again?! The Acme system? Forget about it?! How could I ever forget about it? You never let me forget about it! It was one little indiscretion. I said I was sorry. I was working remotely, I had a couple of beers, this module looked so sharp, and so yeah, I didn't test it. I don't know what I was thinking. I'm sorry, but are you ever going to let it go?"

Every baby is beautiful to its parents. No parent wants to hear someone say that the baby is ugly. Every time a tester finds a defect in the code, the engineers feel like the tester is telling them the baby is ugly. Blood, sweat, and tears went into developing that code, and the nicest thing the tester ever says is that they haven't found a flaw in it...yet. It's a relationship that you, as the project manager, have to manage and set some solid ground rules for or it can turn ugly quickly.

Here are some things that have worked for me:

There Are No Bugs

Don't call a flaw in the code a bug. Bugs are things in nature that crawl into the home and are unavoidable. It's no one's fault that the bug just came in. Call them defects. The code has a defect. Defects are imperfections that the developer left in the code. Using the word *defect* creates a responsibility. Bugs just happen. Defects have owners.

There Is No You in Code

Never use the word "you" when talking about code. Never allow anyone to say, "See here where you check for the permissions, you're not checking the right column." You are talking about code, not the person. Say, "See here where the code is checking for the permissions, it's not checking the right column." This might seem to be a contradiction with my first suggestion, but it's actually not. You want people to be responsible for their code, but you don't ever want them to feel personally criticized. The code has a defect, not the engineer.

Engage Early, Engage Often

Have the testers engage with the engineers early in the project. You need to establish a trust relationship between the testers and the engineers. Have the engineers describe the architecture to the testers. Have the testers ask questions about the interfaces and the timing and the ways things could go wrong. Ask them what they are doing if they hit errors at various points. Error detection and resolution are often *not* covered in requirements. The first time anyone sees the problem is when the tester is going through crazy test cases they've contrived and the system is doing things no one expected…except the engineer. Avoid surprises; have them talk.

Your testers have also seen a lot of ways things can go wrong. Let them share some of these with the engineers and head off problems before they happen. Likewise, the engineers can share points in the system at which all of a given type of activity occurs. This can reduce the work a tester needs to do, since if they test the activity once on one function, then they've tested it everywhere. You get the engineers to help improve the use of your tester time. These exchanges help both sides better understand the end product and how they can help each other get there. If the first time they talk is when the tester has found a problem, there is no reason to believe this relationship will go well.

Encourage Multiple Communication Channels

Have your testers and business analysts and user interface designers get together periodically to review the latest developments and user needs. I love software engineers, but they are not the best communicators and to rely on them to inform the testers of changes is not your best option.

Every change is more work for the testers, and they need as much of a heads-up as you can muster. The changes that might appear small can generate a whole day of added work to the testers, so you have to make sure they know about it.

Establish Rules of Engagement

You need to establish rules of engagement for resolving defects during testing. Interrupting the engineers when you find that something is left versus right justified will irritate everyone. However, if a defect stops all testing in its tracks, it has to be looked at right away. You should establish a vocabulary for the categories of problems. For example, "nits" are things that can be looked at last. They are cosmetic or perhaps minor irritations, but they don't stop work. "Show stoppers" are an emergency that indicate the system is fundamentally flawed, so everything stops until they are resolved. There should be very few show stoppers, ever. "Problems" are errors that need to be addressed soon but can wait for a break point in the developer's day.

The exact language isn't important, but the key is to establish what category of problem is worth interrupting the developers to fix and what can wait for a scheduled or quasi-scheduled time to review. Developers need to be in the zone when they are coding; all those variables and the states of everything are floating around in the air around their heads. Every time you interrupt, all of that tumbles to the floor. Imagine crashing glass onto a marble floor every time you interrupt your engineers, and it will be easier to let them be. They lose not only the time it takes to talk with you, but also the time it takes to pick up all those variables and states off the floor and float them in their proper place in the air again.

Step In Quickly

Intercede only when you need to but do so quickly and decisively when you do. The first and most obvious point at which you will need to intercede is in setting up the proper language. Don't allow the word "bug" and don't let "you" into any conversations. You have to step on that hard. If things are getting tense, these words sometimes creep back in. Don't allow that. I've worked projects during which the relationship got ugly, and in those cases it's often best for the project manager or a business analyst to be the messenger between the testers and the software engineers. Have the testers do all their logging and documenting as they normally would, but then have someone else talk to the developers. "Bob, we're seeing a problem with the reporting function. Let me know when you have a moment and I'll show you what I mean." Then respect their time. If they say they can't look at it now, walk away. Odds are the project is in a crisis/crunch and they can't focus on you when they're slaying some other dragon. If it is an emergency, then say "Bob, we have a show stopper with the reporting function. I need your help to figure out what's going on with it." Bob won't like it, but at least he knows that you've used the magic "show stopper" word and you wouldn't have if it weren't serious.

Be at Your Best at the End

The end of a project or a release is a frenzied period of activity, and this is when you need to be on your toes. To the software engineers, all the testers are doing is dumping on them. The software engineers have been working late nights and weekends, and every time they talk with the testers, it's more bad news and more pointing out of picayune flaws. At some point they will feel picked on. They will feel that the testers are being a negative influence on the project. They might start to let frustration get the better of them and feel like there's no point in testing their own code since that's the tester's job.

The testers have been working night and day, too. They got the "final" code two weeks late; they never heard about a third of the changes that happened in the system; and every time they do a great job, the software engineers dump on them for being OCD. Every effort they make to improve the system is met with derision, and they get zero appreciation. They are making up for weeks of other people's delays, and no one seems to care. The testers will also start getting frustrated at finding the same defects again and again. They found it once; why don't the developers fix it everywhere—take some initiative for a change? The testers will start to feel like they are carrying the load for everyone, and no one cares.

For these reasons you need to be very visible and listen for breaches of engagement rules and protocol. You have to be an ogre when it comes to this. Everyone is tired. You probably are too, but they get to be grumpy and you don't. You can't allow fatigue to ruin the working relationship at the most critical time. If the users or senior management gets wind of a major screaming match between the test team and the developers, they will naturally be distrustful of the system's fitness for deployment. They also won't look favorably on the developers in general. The testers are perceived to be on the user's side, and therefore users will always assume the testers are right.

Testers Are Usually, but Not Always, Right

Sometimes testers are wrong. It doesn't happen often when they have the test criteria written down in front of them, but it can happen. Sometimes they are standing firm on a point of style versus a hard requirement. This most often happens with user interface issues, but it can also crop up with error trapping or error processing. Here again, as the project manager you must step in quickly and publicly make a call on it and take the responsibility off the testers' shoulders and place it on your own. "I hear what you're saying. I get it. I'm deciding that it's okay for the system to display the text in green even though it's an error message. Would you please make a note somewhere that we should revisit this with the users once they have some time with the system?"

Handling Adversity and Typical Challenges

Supporting the Team During Crunch Time

It might have happened, but I honestly can't recall a project in which there was not a crunch at some point in the project. Crunches can be caused by a large number of things:

- Underestimating the level of effort

- Staff who are sick or unavailable

- Last-minute changes to requirements

- A problem that's more complex than was thought

- Deadlines moved up

- Hardware going down

- Delays on related projects

- 1,000+ other reasons

Whatever the reason, the team has to hunker down and get it done within the time allotted. Long hours, high stress, frayed nerves, fatigue, and crabbiness will ensue. As a manager, it's your job to do everything you can to eliminate the possibility of a crunch. You're there to make sure that everyone and everything sticks to the schedule, that the estimates are correct. You have plans in place to quickly handle hardware failures, coordinate with dependent project managers, and so on.

So I guess I've just said I'm a terrible project manager! The fact remains that there are very few times when you can control all those aspects of a project and keep things running exactly on schedule. Things happen that are beyond your control, and you hope to be able to stay within the bounds of the overall project envelope if you slip on any given task. Which almost always means a crunch. Perhaps the user community was a week late getting back to you with a key decision, and now you have one week less to write the code for it. You had a dependency clearly noted in the schedule for this. You highlighted it in every status meeting with the users. You raised it as a risk, but they were late, and they have no desire to give you another week. So it's crunch time.

Many engineers work very well under time pressure. It's an additional challenge and opportunity during which they can demonstrate their intellectual and technical prowess. That's fantastic for the project manager. In some ways the time pressure provides a fixed boundary on the amount of optimizing and "fussing" that can be done with an engineer's code. There just isn't time to keep fussing; they must let the code go sooner than they would have liked. They receive some solace in the fact that they made a ridiculous deadline with a working piece of code.

Crunches also limit the number and frequency of meetings, which engineers hate anyway. They get to roll around in their favorite environment—coding—and are encouraged to do nothing else.

That said, you can't keep your team in crunch mode all the time. Just because you have a quarterback who throws well when he's scrambling away from defenders doesn't mean you'd want to have the offensive line let those guys at your quarterback every pass play. Most of the time, you need to have things move smoothly and calmly and at a predictable pace.

The results of some studies have suggested that a crunch that lasts more than three weeks actually reduces productivity. You get a great surge in the first week as the team charges out to solve the problem. In the second week you're also still ahead, but the increase in productivity is not as great. During the third week the gang is still plugging away but at nearly normal levels of production, despite the added hours, and by the fourth week you are actually behind. The team is fried, and you need to do something different if you want to make a deadline. Keeping the team working crazy hours at this point only results in resentment, lots of defects being introduced into the code, and a surliness you just can't imagine.[5678 9]

I worked for half a year (and shouldn't have stayed that long) at a company that prided itself on building schedules based on a 60-hour week. They had a hero mentality. One of their sayings was, "If you've not had the airplane door close just before you got there, you're spending too much time at the airport." Everything had to be last minute and heroic. I'm not sure theirs was a macho mentality, but in my opinion, it was fundamentally flawed.

---

[5] Ronald E. Jeffries, *Impact of Overtime on Productivity*, http://xprogramming.com/articles/jatsustainablepace/ (April 14, 2006)

[6] Basil Vandegriend, *Overtime Considered Harmful*, http://www.basilv.com/psd/blog/2006/overtime-considered-harmful (January 26, 2006)

[7] Evan Robinson, *Why Crunch Modes Doesn't Work: Six Lessons*, http://www.igda.org/why-crunch-modes-doesnt-work-six-lessons (2005)

[8] Steven McConnell, Rapid Development: Taming Wild Software Schedules, http://www.stevemcconnell.com/rdvolot.htm, (2005)

[9] David Moran, *The Unintended Consequences of Common Productivity Tactics*, http://www.developer.com/mgmt/it-workshop-the-unintended-consequences-of-common-productivity-tactics.html (April 15, 2011)

The first time I reviewed a schedule I built with my supervisor it went like this:

> Supervisor: "What's this? Why are you using 40 hours?"

I was mystified. I answered what I thought the question really was.

> Me: "Well, um, five days at eight hours is 40 hours."

> Supervisor: "No no no, we always use 60 hours."

> Me: "Do we pay them based on 60 hours?"

> Supervisor: "Oh no, but they're used to it."

I was shocked. In essence he said we plan to abuse our employees and have done so for ages. It's our culture.

> Me: "Okay, but the problem is that this thing is not well specified as it is. We are likely to work more than 40, but if I plan on more than 40—especially if I plan for 60—and things start to get ugly, there just isn't any room to go. We'll fry the team."

> Supervisor: "It's not our way; redo this at 60."

Everywhere you went in the building people were pale, wan, and baggy eyed. Every weekend the place was packed with engineers crunching on one project or another and sometimes two. They brought large crowds of graduating computer science majors into the place every spring and after three years of structured abuse would retain about 10% of them (the abused children who still sought their parents' love). With all this pain and suffering, with all this talent, they were no more efficient nor had higher quality results than any other organization I've worked for. They just loved to chew through people.

Stepping off my soapbox about crunches, I have to say that they still do occur, and you will need to adapt your management approach to support the team through them. Here are some tried and true things that I recommend:

Simplify Assignments

During a crunch the team is less able to multitask. You need to simplify your assignments to allow engineers the best possible chance of completing their work quickly and with good quality. If you have a technical star who was supposed to help everyone integrate things as well as complete the hairiest section of the code, you need to pull one or the other off the plate. Either have your star work on the hairiest, trickiest part of the code *or* have your star integrate things. Don't ask him or her to do both; it's not fair.

My recommendation is to give the hairiest part to someone else, even if that person is only marginally up to the task. Have your lead *lead* and continue the coordination among team members. That's actually harder to do, and if it's not done well, you will never finish by your deadline. If you hand off the hairiest part to your second-choice person, the lead will likely still help out but won't be tempted to forget integration and coordination while hip deep in a highly intriguing, highly exciting chunk of the problem. When things are in crunch time, people race to completion. If someone isn't looking closely at everything, you get the railroad tracks not lining up sorts of results. Someone has to make lots of decisions quickly. There are compromises to be assessed and decided upon. You want your lead to take charge of these, while always keeping the overall architecture in mind. If you saddle the lead with code writing at this point, he or she will retreat to the code writing and no one will be properly handling the myriad small issues that arise.

Simplify Functionality

If you are in a hard code crunch, you need to find ways to reduce the functions that have to be developed. At first blush you might think that you can't possibly do that, but actually it's not as hard as you think. In most any function, the vast majority of the code is dealing with and handling errors. If your engineers are any good at all, they are building perfection into those functions. They are planning for full-out network failures, hard drive crashes, crazy stuff being entered by the user, or wild data being returned from the database. I'm not proposing ripping out error handling, but look carefully at the true once-in-a-blue-moon error conditions and decide if you can wait until the next release to incorporate them. See if you can just trap the error, report it to the user, but skip anything else. "The database has been attacked by wild goats. Please try your query again later."

I have seen engineers build in error trapping to prevent users from creating next fiscal year's records while the end-of-year processing is taking place. Although this might seem like a good thing, in reality the new fiscal year's records are created two months after the end-of-year reconciliation *completes*. It's extremely unlikely that these two things would ever happen at the same time. Ask your team if there are some difficult boundary conditions they are working on right now that have a low probability of occurring. Go through each one and make a decision on which can be delayed to another release or ignored like the example above. Doing this violates the engineer's sense of perfection and completeness and also opens up the possibility that sometime in the future another engineer might look at the code and say, "Oh my goodness! There's no provision for wild goats attacking the database! This code is crap!", which is every good engineer's nightmare—that someone will find a flaw in the code, and thus…he or she will be diminished.

So make it clear that you understand what they are saying. That you understand the risk. Thank them for caring this much. Tell them that you know we have to get back to this possibility but that, for now, you're making a call and will accept the responsibility for anything bad that happens as a result. Ask the team to document in the code where they would put in the error handling, but ask them not to code it yet. If there's time at the end, we'll go back and do it. If there isn't time, at least they have indicated in the code that they were smart enough to know wild goats were a potential threat and would have handled it if they had been permitted the time.

Reducing the handling of boundary conditions and unlikely errors is a rich source of reducing functionality without reducing user features. Most of these boundary functions are why it takes 80% of the effort to complete the last 20% of the code. Doing this sometimes isn't enough to get you through to the finish. Sometimes you truly need to cut out user functions and features. If your users have listed a set of reports they want in the system, I argue that roughly 70% of them will never or only rarely be used. When a user is sitting with a blank piece of paper and a wish list mentality, reports get dreamed up that truly have no business value, like, "number of employees by birth month." Unless your company has a birthday cake reimbursement benefit, no one needs this, but I've been asked for sillier reports than that. My old-white-haired-guy metric for reports is five staff days of effort to code, test, review with users, and correct each and every report. Doesn't matter what technical environment I've used—old FORTRAN code, report generators, database add-ons, whatever—it always seems to suck up five days of effort. So this is another rich area for reducing the level of effort.

Depending on your release schedule, it might be prudent to cut out end-of-year or seasonal functions until the next release. This assumes you'll have time before that release to get it done, but it makes no sense to crunch on code today that won't be needed or even exercised for ten months.

Words To Live By: A Report Takes Five Days of Effort

Be Supportive, Stop Talking to Them

If you're a people person or a strong E-type personality, what you need during a crunch is someone giving you energy and confidence. You need people around you, and you need them to talk to you and tell you you're great, gonna get this done, rah rah, siss boom bah! As a manager (and therefore a likely E-type personality), your instincts say it's your job to be that guy for your team. You're going to keep letting them know how much you love them, how great they are. You're going to find the person who looks worn down and go to their desk and cheer them up, because you're a kind and caring manager, and that's how you'll help the team make that deadline. You're a hero!

This is exactly what your average engineer doesn't need. Remember, they find your conversation draining. They are already drained. They need to know that you're impressed with their brains and abilities and effort, that you have confidence in their abilities, and that they are going to meet this challenge. But truly, get away from them and let them do their job. This is a time to let the lead engineer handle details.

Be Supportive, Do Something Useful

If you want to be useful, make sure there are caffeinated beverages on hand. Get snacks and treats and put them in some common area, but *don't* sit there and try to engage them in conversation. Don't ask them to go out for a drink. Focus your efforts on any decisions that the user community needs to make on outside dependencies, on removing roadblocks. Coordinate with the operations team for deployment. Make sure the test team knows what functions you've delayed to a later release or what errors you're not going to fully handle. Talk to the training department about the new screens or reduced functions. Go do a management report. But otherwise leave them alone as much as you can.

On the unusual support front, I once had a team that had been working such crazy hours for so long that they were actually having repetitive motion, back, neck and arm issues. I found a massage therapist who came in and offered them chair massages (clothes on). This allowed them to get coup for working so hard, appreciation that there was still work to be done, and a feeling of teamwork and camaraderie as each of them got better aligned and unkinked. Another time I got a website unblocked for my developer's IP addresses so they could track the World Cup matches. The minute loss of productivity from their checking the scores was well paid back in improved morale and a feeling of special privilege.

Maintain Clarity on the Prioritizations

If you've had some issues with the engineers' priorities, crunch time can be dangerous. This is when they might want to slip back into the old way because they believe they can deliver faster that way. It's their most comfortable mode of operation. You need to help them by continuing to recite the priorities: "We need this to be simple, not necessarily elegant." "We need to stay focused on a simple design, not the most robust." "We need to make sure we handle the stuff users will type in, not the additional functions." "We're willing to sacrifice performance for completeness." Whatever your project has decided, you need to help them stay on target.

Minimize Meetings

One of the classic mistakes (in my opinion) is to increase the number of meetings when in a crisis. The morning standup, the mid-day standup, the end-of-day standup. Management will say these don't have an impact on productivity because they are so brief, but it just isn't so. They are devastating to productivity. Engineers need long, unbroken blocks of time to mentally construct and work through the complex problems they need to solve. If you set up several meetings during the day, you interrupt that thought process, and it takes time to get in the mental "zone" again after an interruption. So don't set up additional meetings, and don't stop by and ask how they are doing when speed is important. Engineers will lose significant time on top of the actual interaction before they get back to the place they were before.

You have to realize that a progress meeting on something that is late is always viewed by your engineers as an opportunity to point out their failure and publicly flog them for it. You are going after the core of their being, their ability to get things done in a brilliant and masterful manner, while at the same time taking them away from getting the work done.

Often there aren't any good answers to the question, "You said we'd be done by now and we're not. How much longer until we're done?" and now you've made them say it out loud and in front of everyone. They thought they'd be done by now, too. They think they are close, or they would have given you a different estimate. Remember, engineers are almost always truthful. They hate lies and liars even more. So they were wrong. They couldn't get it done by their estimate, and thank you very much for making them feel so bad about it. They return to their desk defeated and demoralized.

But it's actually worse than that. They start dreading the meeting an hour—at best 30 minutes—beforehand. They know they're not done, and they know they will have to embarrass themselves in public in just a few minutes. They start focusing on the trauma to come and not on their work. They try to figure out something to say that won't make them appear to have failed. In some desperate times, they will even try to find some way to blame this on someone else. It can lead to some amazing cases of team disintegration, all because management wants to know how it's going. Add to this at least half an hour of angst, depression, and otherwise feeling bad after the meeting is over.

Don't do this to the team. Hold your normal morning standup, and keep the normal format and tempo of that meeting. If management wants other meetings, don't do it. If you can't avoid them, don't have the team go to the meeting; you go instead. Your job is to shield the team from outside disturbances and distractions. You might not be able to answer all the detailed technical questions, but there will be fewer of those if the work gets done, and almost every moment engineers are in a management meeting they are not getting anything useful done.

How to Rein In an Optimizer

As I have mentioned earlier, engineers have a passion not just for finding a solution to a problem but for finding the best of all possible solutions to a problem. They will beam with pride when they have implemented that solution. It is the acme of their day. The problem, and this has become more prevalent when doing Agile development, is that the problem changes. Sometimes the problem changes drastically. The perfect solution usually involves refactoring code, usually several times. As engineers complete a function, they may see an even better way to do it, so they rewrite the code. In doing that rewrite, they see an even *better* way to do it, so they rewrite the code, and so on. This takes time—time you might not have—and even if you did have time, this is all wasted effort that could be better used on other tasks if the problem changes.

You, as the manager, need to optimize your engineer's time. But your optimization and the engineer's optimizations are in conflict. In the engineer's view, you are making them write bad code. There is now an obviously better way to do it, but you're too dim to appreciate the necessity of doing it that way. This drive to find the best solution is extremely powerful, and you would do well to pay it some heed. I've had engineers discover a better way to do a module *after* the code has completed formal acceptance testing. There they were at their desks furiously writing the new method. "Bob, we've delivered the system to the customer already. There isn't time to do this now." But Bob would have none of that. In one instance I even caught an engineer trying to sneak newly optimized code into the delivery baseline...untested. "Well, I wasn't changing what the module did, just how it did it. This code is better. It will still pass all the tests." The logic of this statement is left to the interested test engineer.

There are several common ways that engineers try to optimize their code. One is to optimize its performance for a single task. Perhaps it is the stated task at the moment. At first blush, this sounds good. However, the one thing that never changes is change. Making something very good and very fast at doing one and only one task usually results in a very restrictive and almost brittle piece of code. A greyhound is optimized for running around a track at very high speeds but is no good at pulling sleds across ice floes. The entire module might have to be scrapped and rewritten if it gets too specific to one set of circumstances. "Wow, Bob, I see how that will make the scheduling functions very fast. Now in the next phase we are going to have to handle their payroll functions as well. How would that be handled by the code?" Specialization leads to brittleness.

On the other end of the spectrum you will find engineers optimizing the flexibility of their code. This too sounds great at first, but it can lead to very complex and hard-to-maintain code. I have a Swiss army knife that has a saw blade on it. It does cut wood, but I sure wouldn't want to try to build a house with it. It's hard to hold because of all the many other specialized blades in the knife. Getting the blade out requires opening up some other blades part way and then sneaking a fingernail under the saw tip, and even then it can only cut branches up to about an inch and a half. An actual saw does a much better job. Sometimes it's better to write one module that processes apples and another that processes oranges rather than one that handles all manner of fruit from currants to pineapples. I've had engineers generalize code for highly specialized environments. "Okay, Bob, I get it, but our client is only processing insurance claims documents. They don't do tax returns, wills, or powers of attorney. It's not a good use of your time, and they won't get to use what you're building for them." Generalization can lead to fragile, faulty code.

The first question is how would you even know if someone is busy redoing or optimizing code? It's actually not that hard to figure out. There are certain poker *tells*. For example, Bob showed you how his module worked when you popped by his desk, but now he's working on it again. Another is that Bob was 90% done with something yesterday and is now 80% done. When you ask him about it, he might say something like, "Oh I needed to clean up some of the internals." Or "Yeah, but it wasn't handling all of the possible exceptions." Once they catch on that you're paying attention and trying to make them write bad code, they will start reporting differently.

You need to also listen to the chatter in the cubes. Trust me when I say that engineers are proud of their cleverness, and when they find that new and better way to do something, they will share it with their peers. It's not bragging or boastful. It's joyous. They get so excited about the new way that they have to tell everyone. Imagine if you figured out a way to drive from home to work that saved you 30 minutes. You'd want to share that with your office mates. This is a life changer. This same excitement will wash over the engineer after figuring out the newest, bestest, most optimized method for doing something on your project.

The next question is, how do you rein in this gift without quashing it altogether? After all, you want your engineers to be engineers. You want them to find great solutions to problems. You don't want to tell them you don't care about clever solutions or fast solutions or flexible solutions. The key is to establish what you need optimized right now. You should explain your reasoning in a logical manner. "Bob, this is great, and what I want to do now is show it to our user group. I'm concerned that they are not clear in their own minds on what problem they want to fix or how they want to do business going forward, so before we invest a lot of time and energy into making this up to "Bob standards of excellence," I want them to use it in the lab and come to something close to an agreement. If we make this up to your standards now and then they change their minds, it's going to drive me nuts." Notice that I don't say it's going to drive Bob nuts. He might deny this if it gives him the chance to reoptimize the code. Saying that it will drive you nuts is actually high praise to Bob. You recognize that he is capable of better, that you want to eventually achieve that great stuff he can give you, but that you don't want to waste his time on something that's not ready for prime time.

What you're striving for as a manager is sufficiency. However, sufficiency is not something that most engineers much care for. It is the antichrist of engineering. Good enough for the task at hand is not something that they can feel good about. *Anyone* can do sufficient. It takes a great mind to do optimal. Fortunately, the honest truth of it is that you have more work and more problems than you have time and budget, and this is solely why you need your engineers to get it working well enough to meet the challenge. It is completely 100% true when you say to Bob, "Bob, we have so many other problems that I need your help with that we can't let this one suck up all our time. You've done a great job getting this to work. It works well. It's solid and that's what the customer needed. I need you to hold off making this solid function better and instead take on some of our other challenges."

Dealing with Fights on the Team

I hate drama. I hate drama in my personal life, and I hate it in my professional life. I don't even like drama on TV. I hate the various "reality" shows in which they get people to yell and scream at each other. They cause me stress. Projects will have their drama-filled moments, and it's the project manager's job to mediate it. There are really two types of confrontations that can happen on a project: useful ones and harmful ones.

Useful Confrontations

A loud screaming match can be a wonderful thing for the team. Might not be obvious why, but it's true. When folks finally let loose with all their frustrations and pent-up resentments, it can be cathartic. Everything's out on the table and now we can talk about it. You might have no idea that a pace worker is losing his mind watching a burst worker "goof off." The team might be resentful of a guy who has to carpool home each day at 4 PM while they stay late. Someone might believe their ideas are always ignored and there's someone who doesn't respect them. It will boil and fester beneath the surface until it explodes at some unexpected moment.

Confrontation is healthy and useful but only if it's fair. Fighting fair can be good. Name calling only reveals hostility and resentment; it doesn't tell you what the problem is. As the project manager, it's your job to jump into the fray when the argument gets personal or when the facts are in dispute.

I had a team get all over one guy because he had to pick up his kids at the daycare center by 5 PM every day while they stayed late. It finally came out in a screaming match one day. They started to complain about married people not holding up their end, and that's when I had to jump in and let them know that this new dad was coming in at 6 AM every day to make up for it and that I had approved the work schedule. Since the rest of the gang was getting in around 10 AM and then working to 6 or 7 PM, they had no idea what he was doing. They only saw that he was not there when they were hitting their prime work time after 5 PM. The conversation turned to a bunch of humble apologies and statements of appreciation for what he was doing.

Engineers are likely to get angry and lash out when they feel trapped, out of control, or incapable of doing what they think they should be able to do. One guy completely lost it one day because he thought that everyone viewed his code as defect-ridden. He had had enough and he did not want to hear another word of it. Beet red-faced, fists in the air, he was mad! But it was great for us. None of us thought his code was a mess. In fact we were amazed at what we'd asked him to do and how well it was working. Yes, there were problems with it still, and yes, we were talking about how to fix them, but we were awed by his achievement. His outburst let us do something we'd forgotten to do, which was to let him know how much we admired his achievement. We were all so struck with the opportunity to actually *do* this task that we got completely focused on finishing it off for the big victory. We never said thank you.

If you have a group of people screaming and yelling about what is a good design during a design session, you are actually lucky. You have a group of people who really care about what they are doing. As the project manager, you have to keep the argument from getting out of hand—and, again, never let it get personal—but you should let them argue and let them yell. I will occasionally step in and ask the naïve question to let things simmer down a little if I see people getting too excited. I can play devil's advocate and let the two sides try to convince the simpleton manager of the virtues of one approach over the other. If I see that one person just won't listen to a consensus reached by everyone else, this questioning can be a way of helping the outlier hear the reason for the group's approach. Likewise it can be used to help break "groupthink" when a team has thought themselves into a box and only one person sees it. A facilitated argument particularly with a moron manager in the room can be a wonderful tool.

One thing to watch for, though, is the quiet engineer who is not participating in the design debate but seems to exhibit a look of concern. This person has almost always spotted the flaw but is wary of stepping into the confrontation. When I see this look, I will sometimes interrupt the debate and say something like "Bob, you've been much calmer through this excitement. Do you see anything that we're not already talking about? Do you have any concerns about going down this path or that path?" Now the quiet engineer is talking to me, not the group. I asked his opinion he didn't have to thrust it into the melee.

As a project manager, I have had to learn to quell my personal aversion to confrontation and to instead use the passion to build team cohesion and correct communication misfires.

Harmful Fights

Teams consist of human beings, and human beings are, well, human. They have likes and dislikes. They have histories. They have preferences. They have styles. Sometimes these come together in horrible ways. Sometimes two people will just not like each other. This is not fun. When you have to work week in and week out with someone you don't like or trust, it can get hard on the nerves. Communication is at best bad and usually nonexistent. There will be splendid flourishes of passive-aggressive behavior and the not-so-obvious, left-to-the-interested-coworker-to-figure-out sentence fragment messages. This can get ugly and spread to others on the team. One project I had was like a school playground in that it became all of them against the one outsider. This doesn't help anyone.

You have to step in and put a stop to this kind of fighting as soon as you see it. It's time for you to be the responsible adult in the room. As much as we all hate confrontation and drama, this is a drama-required moment. Sit the sides down together and let them know you've seen what is happening. Let them know what it's doing to the work. Let them know that it's not acceptable. As the project manager, you are not responsible for helping these two to learn to like each other. You're only responsible for enforcing a respectful and professional manner of behavior that is focused on delivering the highest possible quality product to the client. After that, you just don't care.

If you've seen or heard particularly grievous interactions, lay it out there. If you know about communication gone astray, let them have it. Explain that it's not okay to leave out important information in an email. It's not okay to speak ill of one another. When they are in the office, it's about the work, and no one speaks ill of anyone.

It's also important to explain that you need both of them on the project (otherwise you'd have flipped a coin and booted one of them off the team). Point out what you need from both of them and what you need for interactions from both of them. Make sure they know that, for the good of the project, the two of them have to figure out a way to focus on work and not past wrongs.

I mentioned earlier that engineers are rule-based people. They hate it when rules are broken. Most feuds between two engineers occur when one of them feels wronged by the other, usually on a previous project. I was once very excited to have two guys with whom I had worked on different projects before being assigned to a project I was running. Little did I know that they had worked together a couple of years prior, and each felt that they had been screwed over by the other. They had not worked on a project together since, and now it was all coming back to roost. Remember, the reason you don't lie to an engineer is that they will never trust you again because you broke the rule. You'd have thought it was yesterday the way they would coolly glare at each other...but in the no-eye-contact engineer way.

A couple of months into the project I was sitting in my office having it out with them because from where I sat, nothing had been done to anyone and yet they were acting like children. That's when I found out about their past. It's not okay to say, "Well, that was ages ago; let it go!" That won't get you what you're after. Instead I had to let them know that their past was their business, but today we're working on *this* project and on this project neither one of them had wronged the other. That is, except for the disrespectful behavior that was stressing out the entire team. For as long as they were on this project they had to bury the hatchet. We did finish up that project and it was successful, but they still loathe each other to this day.

Good conflict should be guided. Bad conflict has to be stomped on.

Dealing with the Sloppy Engineer

Sometimes you wind up with a sloppy engineer. I'm not talking about their office space or their personal attire. Earlier I covered hoarding and health hazards in the office, and I think I've said all I need to say about that. In this case I'm talking about their work. Sometimes you find yourself working with someone whose code is, to put it gently, subpar. In some cases it's just littered with defects. If the code was supposed to combine first and last name into one field before posting to the database, you find it's only writing the first name or just nulls to the database. You review an entry screen and it's missing a field...or three. The engineer misses that on the first screen of a multipart screen there is no "Previous," only a "Next," yet there is the option on the screen confusing the heck out of the user. They switch fonts or colors randomly. Their code is missing error checking, so the first time they hit a data value out of bounds it all blows up. They might not check to see if the user's search returned no records and just go ahead and display a blank screen. They might not check to see that there's actually enough room to fit all the data on the screen if the user actually entered the maximums allowed for each field. In my experience, there are three major causes of this behavior, and they all have to be handled differently. I discuss each of these below.

The Architect Programmer

I have been blessed to work with many gifted software developers. I've had some who could sit at their desks all day long and crank out code. I've had others who were geniuses at figuring out the overall architecture of a system. It's rare to find both

> "...All his life has he looked away... to the future, to the horizon. Never his mind on where he was. Hmm? What he was doing..."
> Yoda in "*The Empire Strikes Back*"

in the same body. The mind that can see the entire system at once is often not the mind that can see the details through to completion. Architects are constantly thinking about the overall structure. They are wondering if what they're doing now can have any adverse effects on the high-level design they've crafted. Much like the coder who upon completion of a piece of code sees a more perfect way to do it, the architect is also refining and revising the top-level design. The problem is that their constant designing and refining interferes with paying attention to the task at hand. While dreaming and scheming for a more perfect design, the little details can fall through the cracks. Their energy and excitement comes from the clever interfaces between services or from the data structures that can handle any manner of entity. These are the things that will be solid and wonderful from the architect. The nuts and bolts code will suffer...sometimes gravely.

The first thing you have to do as the manager is to recognize that you have an architect. How can you tell? Well, start asking about the code. If your programmer starts telling you about the way the code can handle any situation or could be generalized for use in any application, you might have an architect. If he or she brags about the common interface between the modules and how it is the backbone for the entire system, you might have an architect. If this person goes on about the efficiency of the database structure and query tools, you might have an architect. A nonarchitect is much more likely to show you what the functionality is.

So now you know you've got an architect. That can be a good thing. You need someone to be looking at the big picture and someone who can adapt the big picture when the inevitable big change happens. You need someone who has been thinking about ways to make the design more flexible for a while now and knows how to salvage as much of the design as possible to accommodate the changes. So once you have the person identified (Bob), use him for what he is naturally good at. Task him with the high-level design. If you need him to code (and you usually do), have him work on the infrastructure pieces he so loves. If he has an approach for messaging, get him to lay down the initial high-level code but then, as quickly as you can, take it from him and give it to the detail person you might have on the team. Assign your architect with the next high-level piece. Although this might reinforce a belief that he is only a big-picture guy, the fact is…he is a big-picture guy. You can't get the leopard to change his spots, so leverage it.

A side benefit of having an architect is that more of your team will know the overall workings of the design and know it at a very detailed level. When your architect is under the gun to come up with a way to accommodate that change I talked about, it's the detail engineers who will keep him from making a grievous mistake. "Wait, the message bus doesn't handle binary objects. It only passes the data as text. It doesn't encode them." Crisis averted. It could well be that your architect intended for the message bus to handle binary objects and just assumed that everyone was thinking that, but either way, you have now averted a crushing disaster.

But…what if you already have a system architect? What if that architect is actually better? Well, then you're in for a world of hurt. A fine restaurant has one head chef. Not two or three or more. You will need to turn this *other* architect into a sous chef. This is not easy. With two designers on the team, you have instant competition. There is rarely one and only one way to do things. Each engineer sincerely believes that his or her approach is that one and only one way, but nearly always, they are wrong. Each engineer's approach is the one he or she likes the best. It will have plusses and minuses, but there are a slew of other approaches that could also get the job done.

The project needs people to write code and get it to a reliability level such that it can be deployed and used by the business. Every minute spent debating the approach is a minute you're not getting closer to that deployment. First, you have to make clear who is in charge of the design/architecture. Second, you have to make sure that your sous-chef—let's say, Bob—understands that you appreciate his insights and contributions, but on *this* project, you need him to be a good teammate and just do the coding. Ask him to let you know if there is a fatal flaw in the architecture, but otherwise there should be one voice for the entire team to hear when it comes to design and that voice on *this* project is the other guy's. Ask everyone to come to you whenever they have a great idea for an alternative way to handle things. If they can convince you, the pointy-haired manager, then you can decide to convene a meeting with the architect to talk it through. Don't let your guys just go at it without a referee.

So this solves part of the problem, the definition of roles and an outlet for creative thoughts, but it doesn't get the engineers to make better, less defect-laden code. For that you must appeal to their competitive nature and fear of failure. "Bob, this project is very important to the company. I know you can handle the entire design in your sleep. But for this one I need you to take point on a couple of these pieces that *have* to be right. I need you to get in there and cover every angle of these to make sure they are rock solid. Since everything flows through them, we can't have any defects or unchecked boundary conditions. It's not necessarily the best use of your design skills, but I need someone I can trust to bring this home. Will you do that for me?" The gauntlet is now laid down, and although it will cause them some brain pain to focus on the minutiae, they will do their absolute best to meet the challenge.

The ADD Engineer

Attention deficit disorder (ADD) is a touchy subject. There are loads of opinions on either side of this syndrome. Is it over diagnosed? Do we medicate too much? Does it even exist? I am not a medical doctor, nor do I play one on TV. What I can tell you is that I believe it is real and seems pretty common in the software development world. I work in this software world and not in banking, retail, or any other field, so I have no basis to comment on its frequency rates in other industries. People with ADD tend to have great difficulty focusing on things that don't interest them but are able to sit for hours and hours at a time doing something that does interest them. Some say it is laziness or a lack of discipline, but I disagree. They are trying to work on those other things, but they simply can't do it for long stretches at a time. How it affects writing code is that you wind up with code that has loose ends, unfinished items, and weak error checking.

Identifying the ADD engineer is tricky. Engineers don't talk a lot to begin with, so it's hard to get a handle on what's going on inside their minds. There are certain things to look for. It can be a tell if you ask them about their code because they can't seem to get off of one aspect of it and seem to have spent their entire time on that one part. If they keep saying, "Oh yeah, I meant to get back to that," or "I need to finish this boring part but... ," you might be on to an ADD engineer. A good engineer who can't complete an assignment on time might be dealing with ADD. Some say a sign of ADD is someone who can't sit through meetings, but in a room full of introverts, this is not a reliable indicator. No engineer likes meetings.

If someone shares with you that they are ADD, that's great. However, if you ask the employee, you are asking for a lawsuit, at least in the United States. The good news is that regardless of whether you know or strongly suspect, you'll want to do the same things to help.

- The first thing, and this is not necessarily intuitive, is to assign them a lot to do in a short amount of time. ADD people generally can't work well on a task that isn't in their fascination zone if it's the only thing they have to do. Give them a number of things to do, and each will get nibbled away over time. You also can't leave them with a long time period for the tasks. It has to be pressing. This usually means the tasks have to be smaller than you might assign to someone else—more of an inch-pebble than a milestone. Now, this only works if you also follow through on the rest of this list.

- Help them with the organization of their work. Send them a list of what you want them to work on each day. If you're working in an Agile environment, this is easy. You get to do this every day at the standup. If you're not working in an Agile environment, it is a little more awkward. It's still okay, but you have to spend some time building up trust and making it a routine. If your engineer is feeling picked on, then start making lists for everyone. The little extra effort you spend making lists for people who are already good at organizing their day will pay off in getting your ADD person to hit their potential. Just stop by his or her desk and drop off a list of the things you need for the day. Encourage them to write their own lists as well. But having a list and using a list are not the same thing.

- Depending on the severity of their ADD, you need to pop by and ask how they're doing against their lists. Make it casual. If they have made no progress at all, ask them what other tasks anyone has assigned them. Find out if they're avoiding their own tasks by "helping" someone else on the team. I had one employee I had to check in on every 45 minutes. It was exhausting, but it was worth it.

- Have them take notes in meetings. Your ADD person can be sitting right there in a meeting and hear nothing. Absolutely nothing. His mind is elsewhere. So to help him get the most from the meeting, ask him to take notes. I don't understand the cognitive psychology of it, but it seems that transforming the spoken word to writing and doing something physical (the writing), helps with retention and focus. This isn't perfect. You can sometimes see your scribe drifting off, but this too can be easily corrected with a simple "Bob, did you get that? I think it's important we get that in the notes."

- Have them be the scribe at the whiteboard in design sessions. It forces them to look at the speaker and perform that same spoken word to physical writing memory and focus aid.

- Get someone else's eyes on their work. This is something you should be doing for everyone, but if you have to triage the effort, help your ADD person by getting someone else to peer review their work.

- Some folks say that a quiet work environment is the best thing for an ADD person, and at first thought this makes sense. If you have someone who has a hard time focusing, don't put that person in an environment with a lot of distractions. You should move them away from the cubicle near the loud, clanging door or the coffee pot. Instead, put them in the quietest, most peaceful corner at the end of a dead-end hallway. I have two issues with this thinking. First, for an introverted population, this end-of-the-hallway, quiet, away-from-people cubicle or office is the most coveted piece of real estate. If you give it to your ADD person, there may be some serious resentment. Second, I'm not sure that quiet helps. It's not that they are distracted by the shiny objects that wander into their field of vision. Their mind is hopping from subject to subject all on its own. Creating a quiet, solitary, library carrel–like environment for them will not improve their situation. It's probably a better idea to put your "long talker" off in a solitary area rather than your ADD person. You want the ADD person to hear things in their surroundings that remind them they're supposed to be working on the database interface.

A quick bit of anecdotal, unscientifically gathered evidence. I have a dear friend who is diagnosed with adult ADD. I don't have to guess; he told me. He keeps talking about getting all sorts of things done when his wife is away visiting relatives because the house will be quiet and he will have no distractions. This has never worked. If the world is quiet and no one is asking him questions or the phone doesn't ring, or things aren't clanging around in the background, he can't focus on *anything*. He gets nothing done and then hates himself for his sloth. Instead when his wife is away, he tries to find a loud, busy area to sit with his laptop, and mountains of work come streaming out of him. A coffee shop or bar seems to work well. One of his most successful venues was the concourse of the Venetian Hotel in Las Vegas. Right there by the canal, with all sorts of crowds wandering around, music playing, gondolas rowing by, he was in productivity heaven. I'm very proud to say this suggestion came from my beloved wife, a true managing genius.

The Weaker Engineer

A common reason why someone's code is not up to snuff is that, frankly, they aren't that good. It might be that they are a junior developer right out of school. It could be—horrors!—that they are just plain average...or less than average. Even introverted engineers can be "meh." I mentioned that we all get the same number of points in life. Those for most engineers fall in the analytical, detail orientation side of things and less in the smiley, gregarious people side of things. Some don't get that many assigned to analysis or detail orientation. Maybe they're in culinary skills or sports trivia, but either way you're sitting there with a so-so engineer.

If it's a junior employee, it's a much easier problem to solve. The problem with a recent computer science or information sciences major fresh from school is that all of their assignments only had to work long enough for a demonstration. They never had to be maintained or handle all manner of crazy user behavior. So these recent graduates know some coding skills but nearly nothing about software development. Recall I said that 90% of the coding effort is error handling and recovery from errors. They did about 10% of that in the average college assignment. I am not mocking the college experience; it's just a fact that the average class is intended to demonstrate some basic concepts. You certainly would not want each freshman programming class to maintain the university's primary applications as a learning tool. You need to pair up junior employees with someone who can be their mentor. It's usually not hard to find someone to agree to mentor a new employee. Although engineers are introverted, they do like to share their wisdom and demonstrate their mastery of code and systems lore in a one-on-one setting. You need to check in with them to make sure your new mentor is not being an overbearing ideologue, which has happened, but generally these relationships are a great thing. Your old timer gets to feel smarter than usual, and your rookie learns some good solid ways of doing business.

In the other case, when it's not a junior programmer but a just plain not very good programmer, it's a little harder. You can peer review their work, which you should. You can try to mentor them with someone older and wiser, but frankly if they are just plain not that brilliant, you will likely be settling for higher quality improvements over higher productivity improvements. By that I mean you can usually get them to test their code more thoroughly and to include more errors to trap and handle, but they usually can do this only at the cost of additional hours per task. You will likely have to increase the time allocated to their tasks to account for this. Do not assign them critical path items. Do not assign them particularly complex items. In any system there is a host of work that is not as complex and just requires someone to slog through and get it done.

There might be a tendency to want to evict your so-so developers from your project, but I caution against this. As I just said, a lot of very vanilla work has to be done on any given system. Your stars will become grumpy, annoyed, and resentful if they have to do this type of work. It does not challenge them and therefore it is not enjoyable. To your stars it is tedious, boring, painful, and the worst part of the project. However, your so-so developers are fine doing these sorts of tasks. It's challenging enough yet probably similar to work they've done in the past, so it's not over their heads. Use the skills you have on the team to your best advantage.

The Lazy Programmer

I left this category for last and did not include it as part of the three causes of sloppy code because it's the category of person with which I have the greatest trouble. Sometimes people have defect-ridden, flakey code because they just don't care enough to generate good code. They are lazy. They don't want to invest the energy to produce a quality product. They don't take pride in writing great code. Their interests are elsewhere.

There's a difference between someone who isn't a very good developer and someone who's a lazy developer, the difference being that the lazy developer knows what he or she should be doing but chooses not to make the effort to do it. It's the difference between "Oh wow! I guess I should check for an error on that call," and "Yeah, I thought about that but didn't figure it would ever come up." I can forgive the former but the latter just makes me angry.

Comedian Ron White has an entire routine built on the premise that "You can't fix stupid." I have a belief that you can't fix lazy. Perhaps someone is lazy because they don't get personal satisfaction from doing a good job. They'll do whatever it takes to do as little as possible. As you can probably guess, I have little patience for such people.

I knew a guy with some pretty antique technical skills. I offered him a chance to learn some current languages and tools on a project and made it clear that I had built in time for him to come up to speed. That it wouldn't be a killer death march for him. He told me flat out that he really wasn't interested in doing anything new. He liked the pace he was working now and wanted to stick with it.

One guy asked me if he could start his day at headquarters for an hour and then come down to the customer's office, which is where our work was being done, so that he could charge both the hour checking email and the commuting time on his timecard. His technical work was similarly infused with enthusiasm.

I had an engineer who saw that a data field was named "Accrued_date" and assumed it was of type DATE. She never queried it before she wrote her code, nor did she check the output results. Turns out it was a text field and when mapped to a DATE data type, bad things happened. A simple query to see what the data looked like or a simple checking of that output data from her function would have revealed the faulty assumption. Nope, she just didn't care.

I've had very little luck changing the attitude of lazy programmers. When I had positive results, it was because of praising their good deeds in public and making them feel appreciated. I slowly got a few of them to feel better about their work. But more often than not, my praise was met with resentment and continued bad behavior.

It sounds like anathema to my whole mantra of include everyone and make the best use of all talents, but this is one case where I say punt. Get the lazy developer off your team as quickly as you can. You cannot give them the mundane tasks, because they won't do them well. You can't just add more time to the schedule to allow them to work at their pace. It's not the pace that's the problem. They just don't care. If you keep them on the team, the rest of the team will constantly have to recover from the lazy developer's omissions, sloppiness, and disrespect. They will resent it, and they will turn that on you when nothing changes. It can't be okay to knowingly crank out bad code. It's just plain unfair to everyone else, and they know it. I've been in situations when everyone else was hurrying through their sections to allow time to make up for the next mess the lazy guy was going to create. Needless to say, their own work suffered, and everything became suspect. Get rid of the lazy person.

Telling the Customer Bad News

You have to be honest always, but that doesn't mean you have to cast things in only one light. Nothing is as plain as it appears when it comes to software development. It's still an artistic, individually creative activity. We're not building houses that have absolutely fixed rules for how to build walls, how to install a window, how long it takes to plumb. I'm not denigrating these fields; in fact I'm a huge admirer of anyone who can build these things. What I'm saying is that, relatively speaking, every software project is like trying to build a house with a brand new building material. Or with a completely different type of roof, one that you're not sure how you're going to support. A building in which nails must be heated up and pushed into wood. Nothing is ever the same, even though we're building a house. That's why there are often occasions when we have to go back to the customers and tell them that we're going to be late, it's going to cost more, or it's not possible to do what they asked for in the way they asked us to do it.

Ideally we saw this happening and started talking about it very early. But I suspect we've all been bitten by a slew of things, including denial, that have left us surprising our client. Clients do not have to play by our rules. They can go off the deep end if they want. They get to deny history.

Once I had a situation when I said that doing X would pose the risk of Y and that Y was bad, so we can do X but it should not be done for long or the risk of Y would become greater and greater. When Y happened, they were "shocked." They had no idea it would happen. Why didn't we tell them? What kind of hacks were we? So, we did the obvious thing and showed the weeks and weeks of weekly reports that included Y as a risk and asked for permission to do the right thing to avoid Y. We showed them the decision brief for X in which we said that X needs to be temporary or Y could happen. We showed the monthly reports in which we talked about X and the risk of Y. It didn't matter. They still protested that Y was never made clear to them. So just like in the real world where you can't control other people, you can only control your reactions to them.

Frankly, to a hostile audience, the "I told you this could happen" defense only means that you knew this could happen and did it anyway. Even if you win your point, you are in a losing position.

There are categories of things you have to pass on.

Late Projects and Additional Funding Needed

I'm grouping these together because they almost always go hand in hand. As anyone who has read anything about software projects knows, projects can take longer than estimated. If you have been good and diligent all along the way, the client should know very early on that things are not going according to the original estimate and it's going to take longer. Longer means more hours are needed, and people cost money, so the client should have an idea that more money is also in order. But management has a way of forming a bubble of obliviousness around themselves. It won't matter that you've been telling them for weeks and months that things will be late and cost more; they will be shocked, surprised, angry, dumbfounded, disappointed, and sad.

The best way to handle this is never to deny the situation. There is a tendency of all project managers to believe that something magical will happen and that the project will get back on the original schedule. They believe it in their souls. The facts are there, right in front of us all. For the first half of the project, we've only hit 80% of our target. Unless there was truly something odd about the first half, we're not going to exceed our earlier production rate unless something changes radically, so we're going to be late and we're going to need more money. Those are facts. They aren't even hard-to-gather facts. A quick glance at the schedule will tell you that.

So first, you have to break the news to yourself. Be honest. Let go of all those false beliefs. Ask yourself what would have to change for the team to make up 20% in half of the total time. Odds are there is no way to do this. I recommend posting a to-do task for your weekly schedule to check the production rate against the schedule and force yourself to write down on something visible how far behind you are. Once the melee of the project gets going, you will find ways to avoid looking at this notation. Make it a task for you to complete so it's very hard to intentionally ignore. Once you admit it to yourself, you need to prepare to admit it to the client.

The first step is to gather the facts. The schedule and accounting reports are part of it. Those are evidence of the situation, but they don't explain how you got there. You also need to collect the weekly status reports, in which, ideally, you've been noting the lack of adherence to the schedule, any risks that manifested, outages that occurred, changes to requirements, and the like. *These* are the whys. You also need to do some re-estimating to come up with the answer to the very first question you're going to be asked, namely, "Well, how long and how much *will* it take to finish this?" With these tasks completed, you are ready to speak to your client.

I suggest you ask for a separate meeting from the usual status review. I also suggest you include a small group of the key decision makers and perhaps one or two of their trusted advisers. It is best to call and request this meeting instead of emailing. The conversation has to be quick and frank. "William, I need to talk to you for about 30 minutes to discuss options with you for the project. Looking at where we are now and how far we have to go, we are not going to be able to finish on time and budget. I want to walk you through our status and review some options we have." William will almost always want to get into it right then and there. This conversation should not be take place on the phone if at all possible. If William starts in with a slew of questions, turn it around and ask, "Are you free now? I can come right over. There's a lot I'd like to show you and it's hard to do over the phone." If the client is not free, then set up something for as soon as you can.

There's a nearly standard set of questions a client will want answered:

- **How far behind are we?** The schedule and budget sheets will explain this.

- **How did this happen?** Your review of the status reports, changes, risks, and outages will help explain how. If you're just slower than you expected, that too is something to bring up. "We assumed we'd be able to do this faster. It is more work than we realized (or it's more complicated, or whatever it really is)." Do not sugar coat the problem.

- **Can't we make up the time?** Here's where you have to remain honest with yourself and with the client. No, walk them through the productivity gains that would be required. Explain how unlikely they are to realistically achieve. Remember that the future work also has risks. It's not a downhill glide from here. Also, do not offer to make up x% of it. That is a classic response to the client going through the bargaining phase of their project "grieving" cycle. "Well, I think we could recover about 10 days if we really worked a lot of overtime…" Don't do it. You couldn't make the original productivity estimates, so why would you be able to make higher ones?

- **How come I'm just hearing about this now?** This is another anger response. If it was in the reports, point to the reports. If you told them a week ago, remind them you told them a week ago. It won't help assuage their anger, but it will help assuage their belief that you do know what you're doing as a manager. Do not expect them to agree that you told them. That might suggest that *they* don't know what they are doing or are lazy or incompetent, but you do have to get it out there that you *did* tell them. But what if you didn't tell them before? What if you too were surprised? Well, that's harder. Now you have to explain the nature of the surprise *and* how you're not going to be surprised in the future. You'll probably have to take some lumps for that, but it's always best to be honest about what happened.

- **Well, what do we do about it?** This is a great question. We're moving on to action. Here is where you lay out the options. Same set of work and team, schedule is now this many weeks' longer and costs this much more. Reduce the functionality and the schedule and budget can stay the same, but we deliver less. Add more people and we can hit the schedule, but it will cost more.

I also suggest you ask for a follow-up meeting to brief the client on the new action plan and how it's affecting the status within a week. If they were out of the loop before, you need to bring them back in now.

Faults in the System

I think this is every project manager's nightmare, but we've all been through a couple of truly ugly situations in which the problem is a monstrous fault in the system. These faults seem to cluster around interactions with other systems but can be completely internal as well. They usually result in large-scale data corruption, the kind that requires going back months in the backups to pull up correct data. The problem can take weeks to correct. It might require going back to the legacy system for some period of time until you can fix the new system. These faults can result in data being made public or made available to people who should not have access to them. They are the ugliest of situations, and you now have to explain it to your client.

Once again, the golden rule is to be honest and quick to report the problem. "William, the new system has a flaw that has corrupted the database. We're looking into how much of the data are affected. We don't know that right now, but we suspect it could be all of April's sales data and possibly more. We're also looking at the code to figure out how it happened. The problem didn't turn up in our testing. We're also looking into why it happened, but our first priority is getting the database clean and ensuring it doesn't corrupt any more data."

If they hadn't heard about it earlier, that's great, but odds are they have and they've been catching hell for it. It's important that you say out loud that the number one priority is getting the system up and clean. It is all too common to want to find out how this defect went unnoticed, review the testing procedures, installation procedures, and take all the other process steps, not to mention find the author of the defective module. These all need to be looked at, but none will fix the currently broken system. The forensic work has to be the second tier of actions.

You should be very action focused with your client. "William, here's what's happening now. The DBAs are pulling up the backup tapes so we can correct April's data. They are also going to check other months to see how far the problem might have spread. The developers are going through the code now to find out what chain of events causes the error. They are working with the test team on that. The web team has put up a temporarily out-of-service message on the website. I will have a better feel for what the root cause of the problem is sometime after three this afternoon." In other words, be very factual in explaining the actions. You client will need this information to explain it to the angry users and clients.

There will be pushback on not finding out how this happened and you must be firm. "William, I hear you and we need to figure that out so it can't happen again, but I recommend that we focus our efforts right now on restoring service. Once the fire is out we can review how it happened. In fact we must or we're being foolish."

Major system faults erode good will and confidence in the team, and frankly it can be deserved. I have been in charge of systems that faulted because we blew it, and I've been the innocent victim of someone else's system corrupting my system's data. Either way, you will lose some good will. Your system went awry and caused your manager to lose face. You should expect a cooler-than-usual relationship. You will have to rebuild the manager's trust. This is perfectly normal and not something you should resent. In the coming months your status reports should talk explicitly about how you're addressing the root causes of the specific failure and also list how you're improving things in general to prevent future disasters. Assuming you are not actually incompetent, the trust will return.

Someone They Like Leaving the Project

Clients come to really know a project team, and there will be certain individuals whom they either rely on or just plain like for any number of reasons. Particularly in the consulting world, engineers rotate on and off projects. Sometimes their skills are no longer needed, or perhaps their career needs a new challenge, or their skills need a refresh that they can't get on your project. Regardless, sometimes the person rotating off the project is a person they really love. You have to break this news to them.

This is another time for a private face-to-face meeting. It's another phone call, not an email request for a meeting. "William, I need 15 minutes of your time to talk about a staffing change we need to make. When are you next available?" Ouch…they know it's a big deal now.

If you're rotating someone off the project because the work is done, that's a whole lot easier. The client will miss Bob, but when you explain that we're done with that task and it just makes sense, they will still be anxious and sad, and angry, but they can see the logic. Well, usually. It can also be a purely emotional response. Once a client begged me to keep a person on the contract even though I had nothing for that person to do, just because the client believed that that engineer was the only reason we were being successful.

When someone is leaving the company, it's also pretty easy because you no longer control their schedule and they are leaving *you* as well as the project. Bob has given his notice and he'll be gone in two weeks.

When Bob is leaving for a career opportunity on another project, it can be a bit tricky. In these cases the client will almost always start bargaining for Bob. Well, how about a two-month transition period? How about Bob works half time on my project and half time on that other project? It can be awkward. Odds are they like Bob and wish Bob well. It's best to appeal to their concern for Bob's well-being and his professional development.

Regardless of the reason for Bob's departure, you will need to explain how you are going to deal with his loss. Who is going to fill his shoes? When will the client get to meet that person? Who is this interloper anyway? On a few of my projects the client was so aggrieved by an engineer's departure from the project that they insisted on interviewing all candidate replacements. I refused this one. Aside from the egregious crossing of the line between client and vendor, no one would ever be satisfactory when compared with the client's image of Bob. I had to get the new person in place and show the client that someone else could also do what Bob does.

So you need to explain the transition plan. It should be finite, that is, it must have a firm end date. I've seen clients wheedle and whine and drag out the process well beyond what was good for Bob or the incoming engineer, Betty. Betty could not do her job with Bob hanging around because as long as Bob was there, Betty was ignored by the client.

You need to start promoting the incoming engineer as much as possible. List her by name in the status reports. "Betty has taken over the weekly test reviews from Bob. This has gone very smoothly." The client won't believe it, but you need to keep saying that Betty is up to speed. In some cases it won't be until Betty has performed some heroic deed that they will appreciate her talents. That's not something you can change. I certainly don't suggest creating a disaster for Betty to pull you out of.

Like someone's reaction when a dear friend moves to a new town, the client has a sincere attachment to Bob the person. If they want to have a lunch or happy hour or something to thank Bob for his work and to wish him well, please let them. As much as you're grieving his departure, so are they. Maybe even more so.

Ending a Project

It's taken a while, but the process people have now realized that closing up a project is a big deal and worthy of some serious attention. I'm not going to go through the mechanics of closing out subcontracts, returning equipment and data, filling out profitability statements, and the like. Instead I'd like to talk about the human side of things.

Ending a project is hard. You and the team have been pouring your heart and soul into creating this new system, and now it's installed, it's up and running, people are using it, operators are keeping it alive, the training department has taught at least the first batch of users, and the help desk is handling calls on it. Everything that the business needed you to do is done, and it's now time for you to go away. Emotionally this is hard. You've given birth or raised this new system and *it's* not leaving the nest, *you* are. You need to be aware of your own emotional state as you help the team navigate this process.

For the engineers the end is even tougher. They've had the trials and tribulations of building the system. They've had their shining moment of glory when the users and some senior client person came around and thanked them for all their hard work. But now, there's nothing but uncertainty. They are worried about what they will do next. Who will they be working with? What type of project will it be? They feel out of control of their destiny, and frankly they often are. It is manager types who decide these things and often with no input from the engineer. So often at the end of a project, the engineers are particularly difficult, uncommunicative, and downright grumpy.

The solution is obvious. You need to remove the ambiguity of their future by helping them find the next assignment. If you were paying attention to the "Performance Reviews with the Team" section, you know that you should have already figured out what role or type of project would be best for your team members' individual career goals. If you're on your game, you've already been talking to the staffing

managers or powers that be to learn about upcoming efforts and to
lobby for your folks on those projects. Ideally, you've already secured
them a slot long before your project ended and have been holding the
wolves at bay until you're finished before releasing your folks to their
next adventure.

Sometimes advance planning doesn't work. Sometimes you truly have
no idea what your folks will do next. That's when you have to shift
into high gear and beat the bushes to find them appropriate work
assignments. An introverted engineer is not well suited to expound on
his or her gifts and goals with an extroverted manager. It is your job as
their project manager to advocate for them. You need to call the
managers of the new project to let them know just what your guys can
do for their project and that their project is a perfect fit for your
engineers' career goals so they'll be especially motivated to come
through for them. You also need to share what you've learned about
how to optimize the performance of your engineers. Their
performance on the next project is actually largely dependent on how
you manage their transition to it. If an engineer is someone who needs
a daily list of chores, then tell the prospective manager. If the engineer
is a set and forget kind of person, share that. If the engineer is a burst
worker, make sure the new manager knows this so panic doesn't set in
when he or she is seen walking around.

There's another angle on the project separation angst that the
engineers are going through. They know every line of their code and
wish they could rewrite
it to achieve the next
level of perfection.
They know exactly how
to make it better, but
you would not let them
do it because of time
pressures. It pains them
to think that someone

> My wife ran across a section of
> FORTRAN code she was remediating
> for Y2K with an inline comment to
> the effect of "This will need to be
> changed to a 4-digit year, but doing it
> now will require too much processing
> time over the next 30 years."

will come along behind them some day, open up that code, and mock
it because the obvious optimizations were not done. Some engineers
cannot be pleased about the glory of the completion but instead
perceive only future humiliation when their flaws are laid bare by
another developer. This is a tough one to deal with. It's not unlike the
problem of people with eating disorders who always thinks they're fat

even when they're painfully skinny.

My recommendation is to face it head on. First of all, the team has obviously done a great job to get you to the final release. You should clearly praise their clever and hard work. Let them know that we've proven, through the testing and the initial users, that they've made a wonderfully solid product. You should then ask them to write down, in an outline form, their suggestions for what they might do if they had time to redo it. Documenting the "obvious optimizations" that they would make if the project had time for it takes the sting away from the threat of anyone looking at it later. It's a preemptive "I knew that." Now no one is lesser or stupid or has bad Kung Fu if they have previously stated they would do it differently if time allowed. It won't take them more than about thirty minutes to make an outline, and it will nearly completely alleviate this nagging angst.

"Lessons learned" documents are also a great idea, with the benefits only rarely achieved. Most projects never complete these, even if it's part of the documented project process. Team members have been reassigned to their next projects, and there's no one around and certainly no budget available to write up a document that nearly no one will find, much less read. As depressing a thought as that is, I recommend making the effort to create such a document, not necessarily for the future of the enterprise but for the sanity of the team. Much like the exercise I mentioned in the previous paragraph, the lessons learned sessions allow everyone on the team to vent their frustrations with something that did not go well during the project.

I like the brainstorming session approach, during which one person is the scribe and everyone just gets to list whatever they want. We discuss each item just long enough to clarify what we're suggesting but hold off on a full-fledged analysis until everyone in the room has had a chance to voice concerns. Then we go back through each one and decide what the real lesson was. Lots of times I've found that the lesson has no lesson; it was just something that went wrong on the project, but these are actually great lessons. I know that sounds contradictory. The lesson is that sometimes, even when you're working hard and trying your best, bad things can happen. It's a Bad Things Happen to Good Engineers moment. Not everything that happens has a villain. We lost a week because the server burned up when the cooling system failed. It's not operations' fault. It's not the weather's fault. It just happened. They are not bad people if bad things happen on their assigned task.

The other items—those that we have had a part in—help everyone see that we are always learning. We always gain insight and knowledge from everything we do and everyone we meet. It helps the engineers appreciate that they are not supposed to know everything today or yesterday, that life is a learning journey, and that we now have something we can teach the world.

I like to start the lessons learned session off with something I did poorly and for which I have something we could and should do better next time. I try to set the open and frank tone of the meeting by pointing out my biggest failure on the project. The good news is that the team members will almost always agree with me. The better news is that they will often follow suit by sharing their biggest mistake. Although the textbooks tell you the purpose of sharing lessons learned is to increase corporate knowledge, I think that just so rarely happens. However, this exercise has value in the emotional growth it offers to all team members, and sometimes it gives you—the project manager—the fortitude to stand your ground next time you start a project and see the same risk (like, no dev environment, the wrong staff). It increases your individual knowledge.

Celebrating with Engineers

There are loads of reasons to celebrate on a project, and I hope you take advantage of them. However, it's been my observation that the one celebration that most often happens is the end-of-project fete. So I've placed this chapter here with the suggestion that you not just recognize the completion of the project but celebrate early and celebrate often. Reasons to celebrate? Well, there's the kickoff, the release of the next version, someone's birthday, a new employee, an employee moving to something new, the holidays, or just a night out to blow off steam. But celebrating with I-type personalities is a little different from being with a group of E's whooping it up. If you've been working with engineers for any length of time, then you've been to one of those group happy hours during which all the marketing and sales types, managers, and administrative folks were having a good time laughing it up, telling stories, sharing tales of things in the past, and the engineers came in, had a beer, said nothing to no one, and at some point someone noticed that they had left. What a bunch of losers! Wow, they were no fun. They sure kept to themselves. Suffice it to say this was not their cup of tea.

From an introvert's perspective, standing around in a crowded bar having to yell loudly to be heard with a bunch of people they barely know who seem eager to tell them things they don't need to know is not fun. It is one of the circles of hell. A proper celebration with the team will go a long way to bring the team together and build trust.

If you're celebrating with the team, odds are they've done something wonderful—hence the celebration. I will, for now, assume you're not celebrating the death of the Wicked Witch or some other event that has nothing to do with their accomplishments.

Venue

If you are going to a happy hour, pick a place that has a room you can have to yourselves, where it would be quieter and less crowded. Some bars have nice alcoves that you can stake out and claim early in the evening. Do not pick a trendy club. Do not pick a place that will require fancy attire. Do not pick a place with an overly exotic menu. A lot of engineers I've known enjoy a good beer, and the current trend in microbrewing suits the engineers' personality well. There's a good chance that someone in your crowd is also a home brewer and can share all sorts of interesting tidbits about the brewing process and how the various flavors come to be.

For holiday parties, a hotel conference/ballroom, while not necessarily glamorous, is actually a good choice. The hotel provides all the necessary conveniences in a confined, contained space. Broad, open outdoor bars or seating areas are not as comforting. A large consulting company I used to work at held their holiday party in the Smithsonian Udvar-Hazy Center, an annex to the Air and Space museum. This was brilliant! The space allowed all the Es to have a large open space to mingle and rub elbows and yet provided a set of very techie things for the engineers to walk around and read up on. In general a museum is a fun place for engineers.

Some might think that a Dave and Busters or other place that has food, drink, and lots of video games would be a winner, but it's actually not as appealing. Your game-loving engineers have a pretty impressive set up at home. They have a preferred way to go at each game with their tailored console, so using some lame general population set of controllers in a large open room is no fun at all; it's just annoying. They will likely not have as much success in their play in the arcade area, and that will only embarrass them or otherwise diminish their self-esteem.

Food and Libations

There are some, but not many, engineers who are real wine aficionados. Also, engineers are not always the most adventurous types when it comes to food. They have some strong likes and some stronger dislikes (which they will gladly go on about if you let them) and don't often deviate from their food comfort zone. It's almost always safe to go with a typical bar menu or mainstream American-Chinese menus. Take note of where your crew most often has lunch. If they are bringing back all sorts of wild stuff from food trucks, then you're in luck and have a more sophisticatedly paletted crew than most. If they are always returning with a ham sandwich, then stick to the American menu.

As far as drinks go, holy wars can be fought over Coke versus Pepsi. To the average Joe the "Is Pepsi okay?" question from our server is annoying, because of course it is! However, to a good quality geek this is an insanely stupid question. If you wanted Pepsi, you would have asked for Pepsi. You did not ask for *any* cola beverage they might have lying about. You asked for a Coke! Pepsi is completely different.

I knew a guy who always ordered a half Coke, half Sprite. His blood pressure tripled whenever he got the "Is Pepsi okay?" question. "I'm assuming you're going to try to give me Sierra Mist and not Sprite as well?" he would say to the baffled server. "Um, I don't think we have Sprite." "Never mind! What root beers do you have?" I was always hoping some server would say, "Oh, one moment sir, let me bring you our root beer list," but alas, that has never happened. He was a connoisseur of fine sugary bubbly beverages if ever there was one. If you want to treat your folks right, see which beverages they drink in the office and ask the venue what they serve—the same way you might ask about the house wine for an aficionado. Some engineers care that much about their soda.

For alcoholic beverages, beer is a good bet. Many engineers' spouses and girlfriends might prefer wine, but it's not always necessary to have an extensive bar set up. Hard liquor is not what I've seen most of my engineering friends drink.

Entertainment

Musical tastes of engineers are amazingly broad. Perhaps it's the link between music and math; perhaps it's the spatial reasoning that music is supposed to enhance. Whatever it is, almost every engineer loves music. What type of music? It is amazing how broad the spectrum goes. Engineers often have their headphones on all day while working at their desks, which allows them to enjoy their tunes while simultaneously isolating themselves from the people around them. I know guys who will be listening to Australian metal bands in the morning, Mozart in the afternoon, and finish with Kenny Chesney in the evening. Many of them play one or more instruments, and they have a keen appreciation for music. If you're at a venue where you have an option to have some music playing, do it.

However, although music is much appreciated, dancing is very much unappreciated. Do not ask your engineers to dance. They don't want to. A friend of mine runs a small and very technical consulting company. She had a two-week exchange of emails and voicemails with the head of catering for the hotel at which they were to have their holiday party because the caterer just did not believe that they did not want a dance floor.

> Hotel caterer: "Do you want the dance floor near the front or at the side?"
>
> Technical manager: "No, I don't want it."
>
> Hotel caterer: "So the side then? We can put it anywhere you want."
>
> Technical manager: "No, I don't want it at the front or the side. I don't want it."
>
> Hotel caterer: "So the back then, that's harder but we could make that work."
>
> Technical manager: "No, I don't want a dance floor at all, not in the front, side, or back. Not at all."

Hotel caterer: "It comes with the package you bought. There's no extra charge."

Technical manager: "No, please understand, if my guys find out there's a dance floor, they won't come. There will be no party if there's a dance floor."

Hotel caterer: "How can you have a party without a dance floor?"

Technical manager: "Please, I don't want one. I can't have one."

Movies are great activities for introverted people. Most engineers like science fiction, fantasy, super hero, or action movies. Taking the group to a cinema and draft house, or a matinee of a popular movie, is a good bet. They get out of work early, and they get to sit in the dark watching something they enjoy, eating junk food. It's a winner. If there's an IMAX theater near you, those are popular.

Finally, if you have a space, you can plan the entertainment yourself. Setting up a slew of large screens, bringing in some pizzas, and having a massive Halo or

> **Surprises Happen**
> I was honored to attend the wedding of a very quiet member of our team. Imagine my surprise when the groom nipped out of the reception only to appear moments later in full *Dancing with the Stars* regalia and did a fully choreographed dance routine with his new bride. You could have knocked me over with a feather.

other game night can be fantastic. Everyone gets to show off a little, bring in their game consoles, sample the food, show up the marketing and sales guys, or just play Mortal Kombat. It works.

Ceremonial Speeches

There's a belief that at celebrations someone should make a speech, and I suppose that's partially true. My recommendation is to keep it exceedingly brief. Your team has done some great work, and you should honor that, but keep it short. Everyone wants that recognition. It's a large part of their work-life satisfaction, but they don't want everyone looking at them for long periods of time. If you do not have something positive to say about everyone in the group, do not point out achievements of any individuals. Folks notice if a favorite gets picked out. "Oh, and Jim, great work" does not count as having something positive to say about Jim. It has to be specific, and you have to know what you're talking about. Don't praise Oscar's database design if the team has been railing about the fact that it's unusable. You'll irritate the rest of the team and just encourage Oscar to create unusable designs in the future. You can praise Oscar for all the help he gave to the team to use the database, but even that is sketchy.

You have to be honest and you have to be fair. If one or two people truly took the project onto their backs and brought it home, then yes, everyone knows this, and praising them is fine. But you have to be careful and know that they weren't just grandstanding or being dramatic, because the other guys *do* know and they don't like people getting praise for being a whiner or a fake. When you make your speech, make sure to praise three things:

1. **Their ingenuity:** For example, "Thank you, guys, for coming up with such a great design. This architecture you designed will do great things for the customer. It will carry them well into the future without needing a lot of new development. I'm not sure that any other crew would have come up with this."

2. **Their hard work and dedication:** For example, "Guys, this took an amazing amount of work. Long nights, early mornings, and I thank you for making it happen. Another team might have let the deadline come and go, and I appreciate all you did to make it on the customer's timeline."

3. **Their team work:** For example, "I also want to applaud how you guys pulled together and supported one another. I saw lots of you take on things that needed doing so that others could finish what they were doing even though it wasn't your task. I saw you all solving problems at each others' desks. We're a whole lot better as a group than we are individually, and we're pretty darn good individually."

It's always a good idea to finish up with what the customer is now able to do. What problem is now history; what new, more efficient, or complete process they can operate; or whatever the end-goal of the project achieved. I think it's always a good idea to celebrate accomplishments in the context of the customer's capabilities. The

---

fact that we built a system is moot without this context. The team is here to serve our clients and when we do that well, we celebrate.

Performance Reviews with Engineers

As the manager, you are likely to have responsibility for your team's annual performance reviews. This is a golden moment, but, like the gold idol in the Indiana Jones movies, it is a golden moment surrounded by booby traps. A performance appraisal is a wonderful time to review the year and goals and see how that engineer did against them. It's a time to review the high points and talk through the low points. It's a time to look at the upcoming projects and see what personal goals could be worked on in each of them.

The problem is that you are sitting opposite someone whose personal image is centered around how brilliant and capable he is, and you are going to have to tell him where he is not as brilliant as you need or point out that sometimes his technical breath smells like onions. You have to point out things that may or may not be on his radar as desired skills and try to convince him of their value in meeting his own goals. You have to explain how things he knows, in his heart of hearts, he doesn't do well are things you need him to do better down the road. These are extremely threatening things for engineers to face.

We start with the problem that you're talking to them at all, which, as we've already said, is stressful and draining of energy. So you have forced someone to speak about things that don't compile, with someone who isn't able to do what he does, and you need communicate what areas he needs to work on.

The key is to focus on why engineers manifest the behaviors they do. They understand that things should be optimized. They live this; they get it. They have a painful desire for detailed precision. They have quick minds and do not tolerate fools well, and they think a lot of nonengineers are fools. They care deeply for their work, and they want it to be perfect. They hate it when defects are found in their code. They find it a personal failing. It's much easier to get engineers to hear you if you couch your concerns and areas for improvement in the context of their noble motivation that is spawning it. Below I discuss some common issues with engineers and how to talk them through these issues.

Engineers Who Are Always Fussing with the Code

I've had engineers who couldn't finish their assignments on time because they were always fussing and refactoring their code. This is a common and extremely expensive problem. It leads to code being tested at the last minute, which means that integration time is reduced and the time users have to validate things is reduced as well. Misconnects between developers have to be replaced quickly, so it's not something you can ignore. But how do you tell a developer that he's fussing to no tangible advantage and in fact is reducing the quality of the team's effort?

What has worked for me are comments like the following: "Bob, you are passionate about your work, which is why you are so good, but there are times when that passion is holding you back. I watch you work to make your code as good as it could possibly be, and you refine it when you have insights into how it could be better. Although that's a noble goal, I really need you to help the team with *more* code rather than better code. I need you to turn your code over to the testers and move on to other problems. The testers and a slew of others could take care of any minor tweaks that they might find while you are working on the next block. I can't leverage your talent if you're only working on a small section of code."

"I also need the rest of the team to see what you're doing earlier so that they can adapt what they're doing to the best practices they can find in your code. If you keep working on the code up to the last minute, no one can benefit from the model you're setting. In fact, sometimes they can even miss the basics like the order or arguments, data types, etc. So I need you—and I know this will be hard because you have such high standards for yourself—to settle for something that is solid, let others learn from it, and work on the next section. Do you think you could help me leverage you in that way?"

Now if this guy writes crappy code and none of this is actually true, don't say it. As I said earlier, never lie to an engineer. Never blow smoke up his derriere. He will see right through you. However, in my experience, fussing for perfection is one of the most common reasons developers don't meet their deadlines or always come in at the last minute. Let them off their own hook on that.

Engineers Who Don't Handle Changes Well

Another angle on this same motivation causes developers to fight tooth and nail when changes occur. We've all seen developers who deny the change in requirements and keep the underlying structure of the database or the business logic even when it no longer is valid, solely because they've worked so hard to perfect that model. They keep gluing on data and code around the previous perfection until it collapses under its own weight. They refuse to go back and undo all their hard work, so instead you wind up with crazy code and data that eventually cause all sorts of horrible anomalies, convoluted code no one can maintain, and things the users just can't understand. "Bob, I thought we said that permissions would be the same for any type of document. Why do I have to enter a type of document when I'm adding users who can change it?" Explaining to the users that Bob sincerely believes that it's a mistake to have the same permission model for video and text documents just isn't satisfying to the end users. So how to you help Bob see the light?

"Bob, you show great attention to detail, and you worked very hard to make sure the users got what they truly needed. I can see that you're trying to do right by them and anticipate their future needs. That's great, but sometimes you're too far ahead of the users. Maybe they will need a different permissions model, but that's not what they asked for yet, and even if they might ask for it down the road, we need to keep the code simple so that the maintenance team won't have a problem maintaining it. They might not ever get to this future need, and in the meantime we're asking the maintenance team to get their heads around a more complex system than the users need today. So we have to look after our maintenance team and make it easier on them."

"Adding functions also makes it harder on the testing team since they have to exercise all of those functions even if they aren't in the approved functions set. I need you to tell me when we have these decision points. I need to be in the loop on these "let's invest a little now and reap the rewards later decisions" so that we can make informed decisions and can budget the testing and integration and documentation time appropriately. In theory we're only getting budget to do the approved requirements, so if we go outside of that, we need to get more money and approval."

David A. Oppenheimer

Argumentative Engineers Who Think They're Always Right

Strongly advocating a particular approach is perfectly acceptable. If someone sees that one approach has advantages over another, you want that person to speak up. However, we also know engineers who are constantly arguing to do things their way. They will come up with all sorts of angles on why theirs is the optimal design. They bully, badger, pester, whine, cajole, and just plain wear down the team by insisting on having their way. Some will even get up and take a "my way or the highway" stance. None of this helps. It makes the meetings go on far longer than they should, it demoralizes the rest of the team, and it makes others less willing to offer their own ideas. Always giving in to one person reduces teamwork later on, because others resent that they are always doing it Bob's way. So now you're faced with telling Bob that not every one of his approaches is best. He surely doesn't agree, so it's going to be a fun chat. Unless you're a better engineer than Bob, you don't want to get into a feature-by-feature discussion of the pros and cons of Bob's ideas. That's a dead end. Instead you need to switch it around to help Bob understand his impact on the rest of the team and the project.

"Bob, you have a lot of good ideas and a lot of passion for those ideas. That's great, and I don't want to lose that contribution to our success. But sometimes your passion gets over the top in our meetings. I don't know if you've noticed, but sometimes you're so passionate that you are actually yelling in our meetings. What I'm not sure you're able to see in these moments of passion is that your behavior makes other people uncomfortable. Yelling is never going to convince them that your idea might be the way to go. People get nervous and shut down. So I need you to rein in that passion and keep the discussion to a normal volume."

"But there's a bigger angle on this that will test you even more. People want to feel like they are contributing. You obviously do; it's why you're up there pitching your approach. But so do they. If we do everything your way, how can others contribute? How can they feel vested in the project? If we do everything your way, they are just code monkeys executing your designs—how rewarding is that? These are not stupid people. They have good ideas too, and we need to hear those and use them when appropriate. So what I'm asking of you is to think through other people's ideas and then—and this is the hard part—if there is no serious show-stopping flaw, let's use them. Let's make this everyone's project. Maybe their idea is in a different style than you like, or maybe it's a little more code than you would write yourself, but if it will work, that's usually enough."

"Also, arguing over everything actually diminishes you in the eyes of the team. If you keep knocking down everyone's ideas all the time, no one will listen to you. You're just that guy who hates everyone else's stuff. You'll lose your credibility as an impartial reviewer. You hate everything, so they'll ignore you. Even when you have a majorly important point, they won't hear it. Pick your battles. Speak up when you see a major failure, but see if you can add value to their designs rather than only arguing for your own. I think they will then start to see the insights you have that they can't hear today."

Engineers Who Are Rude to Users

A common rap on engineers is that they are just plain rude to business users. Some of their behavior can be a lack of social graces, but there are times when the engineer's contempt for the business side is completely visible. Frankly, this is unacceptable. It's hard enough to build a system for a business when you have active, engaged users. It's nearly impossible when they won't talk to you because you're just a bunch of horses' asses. It is not okay to let your folks be disrespectful to the users. Users are why the engineers have a job. That said, users have been known to be disrespectful of the engineers, which doesn't help either. Regardless, you're in control of the engineering side, and it's your job to put a stop to it quickly. Engineers often get irritated and then rude with users when users are unable to put their needs, wants, and opinions into actionable words or are unable to describe things at a level of detail of specificity that could be executed. Engineers get frustrated when users don't seem to know their own jobs well enough to explain them.

Some of this frustration comes from the difference between working with humans and working with computers. When you're working with people, you don't necessarily have to specify everything to the nth degree. People figure it out and have common sense and understand the context in which the business is being carried out. The engineers don't have that luxury. As I mentioned before, they are working with, at best, a four-year-old intellect that forgets everything you might have taught them yesterday every single day. Business users working with people don't even recognize the decision points sometimes. They've been doing their jobs so long it's like muscle memory. They're not used to explaining how to do their jobs to a four year old.

Other ways that business users frustrate engineers is in their lack of vision. This sounds like I'm taking a dig at the users, but that's not my intent. The business users do the business. It's their job. When a bunch of engineers come around and ask them loads of questions about their work so some of it can be automated, most often the users are taking a literal viewpoint of what that means. If we used to get a request from the customer, filled in a form, and then sent it to department X for approval in the past, the users see a future automated world in which they fill in the form and a computer sends it to department X. Makes sense, no? But the engineer sees a world in which no one fills in the form at all. The data are already there; they were entered by the customer already. Why would one need a form? The data would go from the customer straight through to the approval.

This disjoint combined with less-than-superb communications skills can lead to conversations like the following:

> Business user: "So then the system needs to give me a form so I can fill in the data and send it to Department X for approval."
>
> Engineer: "No, you don't need a form."
>
> Business user: "Yes, yes I do. I have to get approval from Department X, and they sign the form to do that."
>
> Engineer: "You don't need a form to do that. The data are already there."
>
> Business user: "No, they're not. The data go on the form so Department X can approve it."
>
> Engineer: "The form is irrelevant; you don't need a form. Forms are stupid."

And it just gets worse from there. At some point the business user is going to talk to a supervisor, who is going to talk to *your* supervisor, and you're going to have a big meeting in which you'll have to explain why you refuse to build the system the business users need.

But this section is about how to coach your engineers on not being frustrated with the business users' lack of vision.

"Bob, I've noticed that sometimes you seem very frustrated with the business users. There were a couple of meetings last month when they got you so upset that you were actually rude to them. That's not good for anybody. I noticed that they most often got to you when you were trying to explain how the new system would work. You need to keep in mind that you got hired to think about how to do things better and more efficiently. You're here to make things different. They are here to do things the old way. They weren't hired because they could envision a completely new world. They're here to work the world they have today. "

"You're better than they are at visionary thinking, so you need to help them see what could be. Don't be too hard on them. They're doing the job that keeps this whole company going, and I sure as heck don't want to do that job! They don't know computers except as a web browser. They've not seen automated forms processing software before and therefore couldn't imagine what the new system might mean for them. They've only seen the world they have—the paper-based one. So what I need you to do is to remember where they're coming from and gently lead them to the Promised Land. Explain that once the customer enters in the request, the data are available for any process we might need. They're now in the computer and don't necessarily have to be transcribed again onto a form. Don't tell them they can't have a form; tell them they might not need it. Let *them* come to the conclusion that they don't need the form. They'll get it if you explain it to them slowly. You need to teach them what is possible."

Engineers Who Are Rude to Other IT Departments

Have you ever received a call from the head of the operations team telling you that they don't ever want to hear from Bob again and that little piss ant can go...? Most technical managers have. This is not good for any project. As I explained in earlier chapters, the operations folks are your lifeline. They control more than you do, and pissing them off is never a good thing. Bob probably was insanely rude to them. There's no point trying to defend his remarks as you likely have no leg to stand on. Operations might have been rude to Bob, but trying to win a "who was ruder" or "who was rude first" debate really won't solve anything. You can't have your guys mouth off to anyone, but there does seem to be a constant tension between developers and operations. Some of it is because both of them consist of engineers. Neither group was gifted with an excess of social graces, so it's easy to see this rudeness devolving into a yelling match with hurt feelings on both sides.

Almost always the argument is about what to optimize. The development engineer wants to optimize the ease with which the new system can access resources, including database, disk space, network, and other supporting applications. The operations guy needs to make sure that they are fully compliant with all the security and governance rules and has to allocate resources sparingly to support all the applications in the enterprise, not just yours. As a rule, the developers are always asking for the world, but the operations team is only willing and able to give them a small atoll in the Bering Strait. But you can't have your guys making trouble, so what do you do?

"Bob, you and the operations team are really at odds these days. This hasn't been a positive relationship." You will likely get a long list of reasons why they are stupid, bad at their job, and should all be fired. How they are not needed, how Bob could manage the server, and a lot more anger and frustration venting. When he gets winded you can begin again. "Bob, I love how much you care about this project being successful, but sometimes I see that getting the best of you. The operations crew has a lot of demands on their plate that I don't think you'd want to deal with full time. They have responsibility for auditing compliance with the security requirements, managing the records management plan, keeping all the servers patched, managing the hardware maintenance, and answering calls from all sorts of projects, ours included. I think we want them to do those things because frankly none of us wants to do them. They are short staffed as it is, and it's not likely to change."

"They really don't have a lot of time to handle new requests, like when we ask for a new development environment. So when we want them to do something not already on their very packed schedule, we need to cut them some slack. They are doing us a favor. If we followed the full process of requesting and reviewing and approving and then scheduling, it would take months to get even the simplest stuff done. They also have to support the entire enterprise, and they have a number of sometimes reliable servers that are stretched pretty thin as it is. When they don't give us an entire disk array for testing, it's not because they're keeping it for themselves. They probably just plain don't have it available. They have to keep everything up and running, so that means we don't always get what is easiest for us. That's very frustrating and puts a greater burden on us than would otherwise be there, but it's just the way things are. So if you are rude to them about things they can't change, they are not as likely to help us out in the future. I need you to learn a little more about their job and appreciate what they already do for us, even if it's not everything we think we need."

How to Structure the Review Discussion

The agenda for a review meeting is worth considering. Many folks like to go over all the accomplishments and achievements first and then hit areas for improvement. I recommend against that. You're ending on a bummer note, and honestly the engineers can't appreciate the praise you want to give them if they are sitting there cringing, waiting for the bad stuff.

I prefer to talk about the projects they worked on and how they felt things went in chronological order. Make it a review of the year. See if you can get them to talk about what they wanted to get out of the project back in March and whether they succeeded at meeting that goal. Ask them about the problems they encountered and what they learned in overcoming them. Ask them about how things went with the other members of the team, the users, and the in-house organizations. See if you can get them to raise some of the issues I've mentioned above. Ask them who they think they work with best and why, and who do they have difficulty working with and why. Although technical coaching is important, it's human skills coaching that is most often overlooked and holds engineers back from reaching their full potential. Spend your time on it.

Once we've been through the projects, I like to move on to plans for next year. Odds are they are on a project at the moment, and you need to see what their goals are for that one. What do they hope to learn from it? What skills are they going to work on? I like to move on to future projects, but before that, I recommend talking about long-term goals. Let's say someone tells you that they want to stop being a Java developer. Their real dream is to be a DBA. That's great. To be a DBA in your company, they'll need to get a certification and that will mean attending some training. Find out if they learn best in a classroom or online at their own pace. They'll also need to spend some time learning from the current DBAs. Think about projects where they could still leverage their Java skills while learning from the current DBAs. Work together to build the transition plan from today's Java developer to tomorrow's DBA.

Whatever their goal, it's incumbent on you to point out what skills they either need to learn or need to improve to make that goal a reality. You should help them understand that it might take a while but you'll work with them to make it happen. Hardest of all is to get an earnest engineer to understand that sometimes they just need more experience to take that next step up (wherever up is). Engineers are great at concepts. Once they learn the concept, they assume they are ready to go. The real world needs concept, execution, and that which no amount of precociousness can replace: wisdom. Wisdom comes from experience.

If there is any takeaway from this chapter, it's that you can't be a good manager if you are not willing to tell your folks the truth about themselves. They need you to help them see themselves as valuable performers, warts and all. If all you do is tell them they did a good job, thank you very much and please do more, oh by the way, here's your raise amount, then you're just filling a chair and they know it. Know your people. Know where they want to go, and help them make and carry out a plan to get there.

Rewarding Engineers

Celebrations are a form of reward, but in a business setting folks often expect and arguably deserve more. Most engineers are well paid. However, money is not their primary motivation. Money feels good for a brief time, but a bonus will not carry that engineer's mood for long. The salary you pay your engineers is important. The good ones do cost more than the bad ones, but to the engineer the salary is a reflection of how much you value them more than what the salary can do for their lifestyle. Don't get me wrong, every engineer has the fantasy of being the next Steve Jobs, Bill Gates, Mark Zuckerberg, or Sergey Brin—the guy who creates something massively popular and then has more money than god and can buy any new toy on a wish list. But it's the first part that is the greater accomplishment in their minds: creating something massively appreciated for its innovation, brilliance, speed, what have you. That is the goal of an engineer. I know some non-engineers who want the pot of money and don't care if they win it or create it. They just want the money. Engineers want the recognition and love of their creation first. The money would be pretty cool too but not first in their minds.[10]

If you want to reward engineers with something that will make them happy longer than money (but not in lieu of money), promote them. Give them the next higher badge and title. It's a public statement that Bob is better than he was recognized as being before. If your firm does not use titles, then make sure that everyone knows that Bob is the guy you're going to rely on when the next tough challenge comes along, that Bob is your "go-to" guy. Ask Bob to take on mentoring some of the more junior engineers (assuming Bob has the communications skills to do that). Put Bob in a position to run code reviews or design reviews. Pull Bob into a review role for other projects that he's not actually on. Leverage his talent to help out more than one project. Bob will appreciate your confidence in him more than a $100 gift card could ever say thank you.

Toys and technology are another good way to make the team feel loved. It's hard to get an engineer toys. It's usually what they spend their discretionary cash on anyway, but there may be things that they would not buy for themselves. Back in the days when PCs were still pretty pricey, we got one of our engineers an off-brand PC. He was thrilled! He almost wept. He couldn't believe it! It was like delivering a pony to a little girl. For months he kept thanking us for his gift, and for years he told us how he had upgraded it or what he had done on it.

For a while Lego Mindstorm robot kits were good bonus gifts. Electronic tablets for handwriting or art were also popular. We got one guy an Apple Newton that he is still wearing on his hip today. Sure, it's now an anachronism, but he is still getting coup for owning it.

---

[10] Dan Pink, Drive:The surprising Truth of What Motivates Us, http://www.youtube.com/watch?v=u6XAPnuFjJc, (April 1, 2010)

Timing might never be on your side, but video games are good choices. The problem is that the engineer will have preordered the title months in advance, but if you're lucky you might be able to use this. Wireless devices are still coming out all the time. Home automation systems that let your engineers wire up their coffee pots to the Internet are winners.

There may also be toys that have nothing to do with technology that would be appreciated. For one guy who loved to play guitar, we bought a Midi interface (back when they were new) and software that would transcribe his played music to written music. He loved it. My boss once gave me a check and said, "We wanted to thank you for everything you've done by getting you a table saw, but we were afraid to go out and get it, since we knew you'd have something specific in mind, so please take this check and get the saw you want." I nearly wept, and I still love that saw, not only for what it does, but also for the love and appreciation that seeing it reminds me of every time.

You might need to work with your HR department to depart from the corporate-approved gift card regimen. In our litigious society getting someone something that wasn't on an approved list might cause problems, but I strongly urge you to tailor your rewards to the individual and to look at their interests and hobbies outside of work as inspiration for those gifts. It says you care not only for their interests in the office but for them as a more complete person. It makes them feel not just appreciated but understood and valued.

What People Taught Me

I've been blessed to have some truly wonderful people come into my life. My mentors were amazing people, and I can recall those moments when I was just stunned at their wisdom. I so wanted to be that wise and clear. I strive for that to this day. Many of my mentors tried to teach me the things listed below, but certain points finally sank home from one individual. Somehow, this person was the one who got it through my thick skull, who gave me the "aha!" moment. These gifts are precious, and although "thank you" doesn't really suffice, it is the word we have. I thank each of them for these gifts.

How could I possibly summarize or reduce to a couple of bullet points the hundreds of hours of learning and the wealth of insights I was given by each of them? I can't really. But just because I can't doesn't mean that I shouldn't try. Thinking back to times when I was passing on something, I asked myself, "Whose voice am I hearing instead of my own?" These are in a rough chronological order:

Alfred G. "Waboos" Hare

- Love everyone, warts and all.

- You know what's right, do it.

- Helping the other fellow is its own reward.

Captain Marshall Sherman, USN

- The Knights of the Round Table never left us. They wear a uniform and serve our country today.

Dr. Michael Duffy

- Fewer words, well chosen, will better make your point.

- Accept that what you know might be wrong.

Ed McGushin

- The difference between managing people and leading people is all about the people.

- Working hard and playing hard can be the same thing.

- Set the bar very high, and then do everything possible to help your people over it.

- Love your customers, for without them you have no purpose.

- Love your life. Live it large, live it loud, live it with gusto, and live it bravely.

- You can do anything that is not legally forbidden.

Dr. Rolland Fisher

- If you can lay out the vision, others will joyfully fill in the details.

- You should work smarter, not harder, but most times you have to do both.

- There is a great deal of wisdom that you can convey with the right saying.

Stan Lucas

- Give an engineer the problem and let the solution be their own.

- Control needs to be given before people can invest themselves in their work.

- All glory goes to the engineers. All shame to the manager.

- Let the customer discover the solution you already see. Then it's theirs.

Dr. Ernst Volgenau

- In business, and in life, honesty is the highest virtue.

- Some things are harder for you than others. You will have to work harder on them.

Barbara Sada

- You control whether to accept ownership of someone else's problems.

- You don't have to be loud to lead.

Dr. Hatte Blejer

- Genius comes in odd packages. Don't tolerate them, celebrate them.

- Learning should never end.

Katharine Murphy

- Strong, powerful, visionary leaders need not conduct themselves like men.

George Nicolau, my stepfather

- Nothing got settled while yelling.

And last but also first, my father, David E. Oppenheimer

- There is no limit to what you can accomplish if you don't care who gets the credit.

About the Author
David A. Oppenheimer is a technical management consultant, author, and speaker. He has worked in the software field for over 30 years building systems for commercial and government clients, as well as leading the technology teams of two start-ups. He holds a Bachelors degree in Physics from Hampshire College, a Masters in Engineering Management from the George Washington University, and a Project Management Professional Certification from the Project Management Institute. He lives in Fairfax, Virginia with his wife, Elise (also a Nerd Herder) and is the proud father of Megan who, despite her best intentions, is also in the technology field.

48077197R00157

Made in the USA
Columbia, SC
05 January 2019